Uncertainty-driven Innovation

"This book offers an original and useful contribution to the field of innovation management. It sheds the light on the benefits and the drawbacks related to innovation in a high uncertainty world, with a particular focus on manufacturing companies."

—Sara Sassetti, *Organization Studies, Department of Political Science, University of Pisa*

Giacomo Marzi

Uncertainty-driven Innovation

Managing the New Product Development Processes
in an Unpredictable Environment

Giacomo Marzi
Department of Management, Mathematics and Statistics
University of Trieste
Trieste, Italy

ISBN 978-3-030-99536-2   ISBN 978-3-030-99534-8   (eBook)
https://doi.org/10.1007/978-3-030-99534-8

This Palgrave Macmillan imprint is published by the registered company Springer Nature Switzerland AG.
The registered company address is: Gewerbestrasse 11, 6330 Cham, Switzerland

Contents

About the Author

Giacomo Marzi is Assistant Professor (Tenured) of Management in the Department of Management, Mathematics and Statistics (DEAMS), University of Trieste (Italy). Previously he was Senior Lecturer in Strategy and Enterprise in the Department of Management, University of Lincoln (UK), where he now holds a visiting fellow position. He holds a PhD in Management from the School of Economics and Management, University of Pisa, Italy. His primary research fields are innovation management, new product development, strategic management and entrepreneurship. He is author of three books and several papers that have appeared in journals such as *Technovation, Journal of Business Research, IEEE Transactions on Engineering Management, Human Resource Management Journal, International Journal of Production Research* and *Scientometrics* among the others. His work has also been presented at various conferences including Academy of Management, European Academy of Management, British Academy of Management, IEEE TEMSCON and Innovation and Product Development Management Conference.

LIST OF FIGURES

LIST OF TABLES

CHAPTER 1

The Unpredictable Environment and Role of Innovation

Abstract Nowadays, companies are facing extremely unpredictable and uncertain environment. Innovation has always been recognised as competitive leverage, but now it has become more essential than ever. By innovating and developing new products and services, companies can gain and maintain a sustainable, competitive position. However, innovating in an uncertain world is not simple. Taking these premises as a starting point, the central goal of this book is to empirically explore the role of environmental ambiguity and instability in influencing the new product development process, product innovation and process innovation, focusing particularly on manufacturing companies.

Keywords Innovation • Unpredictability • New product development • Entrepreneurship

The father of economic theory, Adam Smith, emphasised in his seminal and well-known essay *An Inquiry into the Nature and Causes of the Wealth of Nations* (1791) that while innovation imposes capital expenditure, it is a critical economic practice for promoting wealth. Although the role of innovation was recognised in the sixteenth century, it took another two centuries for Schumpeter (1934) to offer a systematic theory of innovation (Pavitt, 1998).

© The Author(s), under exclusive license to Springer Nature Switzerland AG 2022
G. Marzi, *Uncertainty-driven Innovation*,
https://doi.org/10.1007/978-3-030-99534-8_1

Schumpeter emphasised the importance of economic forces in technological development, stressing the role of innovation as a necessary and chief engine of economic growth, with entrepreneurship at its pinnacle. Schumpeter made a distinction between a manager and an entrepreneur; an entrepreneur sees a gap in the market and fills it with a product while a manager solely organises the labour, the operation and the business (Brouwer, 2002).

In the opinion of Schumpeter (1934), economic expansion is characterised by "creative destruction" with existing monopolies just being provisional/short term because of newcomers' "catching up". In an ideal economy where innovation thrives, replication will have a significant impact on earnings, bringing them back to normal levels. As a result, Schumpeter came to the realisation that perfect rivalry and entrepreneurship cannot coexist.

Consequently, Schumpeter's initial hypothesis served as the foundation for subsequent scientific economic literature that used the concept of innovation as a development engine. Currently, a lot of scientific research on innovation exists, and this is a topic that businesses must address if their aim is to build and sustain a competitive edge and/or get hold of new markets (Marzi et al., 2017). Among many, this is the most vital element that influences an economic system's external competitiveness, growth, production and employment (Brouwer, 2002; Hospers, 2005).

While Schumpeter clearly conceptualised innovation in the firm's context and defined its scope as a process, product and business model, there is still a discussion about many characteristics of innovation. For instance, innovation requirement and intentionality, sufficiency, favourable nature, successful implementation and diffusion could altogether provide a more precise definition of it.

Consequently, a concept of innovation was developed by the Organisation for Economic Co-operation and Development—OECD (2018), which includes all scientific, technological, operational, financial and commercial activities that are required for developing, introducing and marketing the products and processes that are new or improved.

As defined by the OECD (2018), product innovation can be seen as a novel or considerably enhanced good or service. Examples include significant advancements in technological requirements, parts and products, product software, user-friendliness and other practical features. In contrast, process innovation is defined as a new or substantially upgraded processing or distribution system. Significant enhancements to procedures, facilities and/or applications are all part of this (OECD, 2018).

Nevertheless, innovation is a concept that is broad and has many facets, and all its potential layers in the above description are not covered (Marzi et al., 2017). A more detailed concept of innovation proposed by scholars refers to producing or introducing, assimilating and utilising something new with an added value in social and economic systems. It also includes the regeneration and extension of goods, markets and facilities; the creation of alternative production techniques; and the creation of alternative systems for management (Hospers, 2005).

This extended concept captures a variety of essential innovation aspects: it incorporates both internally developed and externally embraced innovation; it emphasises innovating as beyond just an activity that requires creativity. It emphasises expected advantages: the option that innovating can be associated with its own relative novelty is left open, and finally, it emphasises the functions of creating value for the actors involved in the innovation process (Marzi et al., 2021).

As previously stated, the concept of innovation has developed over time, becoming a broader umbrella that encompasses a diverse range of phenomena from various fields of study, while some management authors re-emphasise the significance of innovation in a company's success and viability in the long term. As a result, there is a broad agreement that innovating products and processes assist businesses in gaining a competitive advantage (Marzi et al., 2017).

Therefore, the goal of the present book is to highlight a variety of research streams within the field of innovation studies, with a focus on product innovation, process innovation, New Product Development and uncertainty-related issues about innovation.

The present text represents a comprehensive and extensive analysis of topics that have been largely investigated by the author, who has contributed to the scientific debate with a series of theoretical and empirical studies over the years (Bianchi et al., 2019, 2020; Marzi, 2018; Marzi et al., 2017, 2018, 2021; Marzi & Caputo, 2019). Therefore, this book provides a compendium as well as an inclusive perspective on several topics related to innovation management, entrepreneurship and innovation, innovating processes and products, and the creation of a novel product that is analysed under the light of uncertainty-driven innovation (Rialti & Marzi, 2020).

Chapter 2 of the book is devoted to the exploration of entrepreneurship and the role of the entrepreneur as the fuel of innovation, their ability to create knowledge and the need to navigate the unpredictability of the external environment.

Next, Chap. 3 presents a comprehensive and up-to-date round-up of the research about product and process innovation, with a specific focus on manufacturing firms. Manufacturing firms have been selected because they often struggle with proposing new products, especially when it comes to small and medium-sized enterprises (SMEs). Chapter 3 also proposes case studies of the successful application of innovating processes and products in a series of manufacturing SMEs.

Chapter 4 focuses on some detrimental effects of uncertainty in the development of a new product, namely "Over Featuring" defined by Marzi (2022) as a set of very dangerous but hard-to-escape tendencies that bring to the product and service development more (excessively) of what the organisation resources, the users, the market and the plans actually need.

In continuation, Chap. 5 proposes an empirical investigation of the function of uncertainty, the strategies and tools to mitigate ambiguity throughout the process of developing a new product.

Based on the findings in the two preceding chapters, Chap. 6 further extends the exploration of the role of different forms of "Over Featuring" on the various stages and approaches for New Product Development process Marzi (2022).

Finally, Chap. 7 proposes a summary of the present book, concluding remarks and a series of reflections on further development of the topics covered herein.

References

Bianchi, M., Marzi, G., & Guerini, M. (2020). Agile, stage-gate and their combination: Exploring how they relate to performance in software development. *Journal of Business Research, 110,* 538–553.

Bianchi, M., Marzi, G., Zollo, L., & Patrucco, A. (2019). Developing software beyond customer needs and plans: An exploratory study of its forms and individual-level drivers. *International Journal of Production Research, 57*(22), 7189–7208.

Brouwer, M. T. (2002). Weber, Schumpeter and knight on entrepreneurship and economic development. *Journal of Evolutionary Economics, 12*(1), 83–105.

Hospers, G. J. (2005). Joseph Schumpeter and his legacy in innovation studies. *Knowledge, Technology & Policy, 18*(3), 20–37.

Marzi, G. (2018). *Product and process innovation: From manufacturing to IT firms.* Doctoral dissertation, University of Pisa.

Marzi, G. (2022). On the nature, origins and outcomes of Over Featuring in the new product development process. *Journal of Engineering and Technology Management, 64,* 101–685.

Marzi, G., & Caputo, A. (2019). *Responsible entrepreneurship education: Emerging research and opportunities.* IGI Global.

Marzi, G., Ciampi, F., Dalli, D., & Dabic, M. (2021). New product development during the last ten years: The ongoing debate and future avenues. *IEEE Transactions on Engineering Management, 68*(1), 330–344.

Marzi, G., Dabić, M., Daim, T., & Garces, E. (2017). Product and process innovation in manufacturing firms: A 30-year bibliometric analysis. *Scientometrics, 113*(2), 673–704.

Marzi, G., Zollo, L., Boccardi, A., & Ciappei, C. (2018). Additive manufacturing in SMEs: Empirical evidences from Italy. *International Journal of Innovation and Technology Management, 15*(01), 1850007.

OECD/Eurostat. (2018). Oslo manual 2018: Guidelines for collecting, reporting and using data on innovation. In *The measurement of scientific, technological and innovation activities* (4th ed.). OECD Publishing/Eurostat.

Pavitt, K. (1998). Technologies, products and organization in the innovating firm: What Adam Smith tells us and Joseph Schumpeter doesn't. *Industrial and Corporate Change, 7*(3), 433–452.

Rialti, R., & Marzi, G. (2020). *Ambidextrous organizations in the big data era: The role of information systems.* Palgrave Pivot.

Schumpeter, J. A. (1934). *Theory of economic development.* Harvard University Press.

Smith, A. (1791). *An inquiry into the nature and causes of the wealth of nations* (Vol. 1). Librito Mondi.

CHAPTER 2

The Entrepreneurs, the Innovation Process and the Management of Uncertainty

Abstract The entrepreneurs are the primary drivers of innovation, while the entrepreneurial system is the pinnacle of decision-making and the centre of the innovation network. The entrepreneurial system is not only limited to setting objectives, establishing priorities, interpreting and ruling the reality of a dynamic market, but it is also composing, deconstructing and recomposing the value creation process. The entrepreneur is the one who models and arranges the output based on intuitions and judgements. Innovation is one essential component of the engine of entrepreneurship. This chapter presents the key concepts of entrepreneurship by establishing a connection between entrepreneurship and innovation.

Keywords Entrepreneurship • Innovation • Creativity • Knowledge • Value • Uncertainty

2.1 The Entrepreneurial Structure in the Literature

The role of the entrepreneur in animating and directing innovation processes is critical for a business: their personalities, expertise, proclivity to take chances and ability to look beyond for tools and skills that they lack for the advancement of new processes and, consequently, the company's progress (Ciappei, 2005; Pellegrini et al., 2021). Schumpeter (1934) saw

© The Author(s), under exclusive license to Springer Nature
Switzerland AG 2022
G. Marzi, *Uncertainty-driven Innovation*,
https://doi.org/10.1007/978-3-030-99534-8_2

scientific progress as the inevitable result of a loop. The entrepreneur who innovates is the soul of capitalism and acts as a spark for development in this cyclical paradigm, which the economist describes as creative destruction. Inventions can be turned into innovations because of the dynamic efforts of entrepreneurs and innovators. The entrepreneur is expected to have a distinct personality and an openness to creativity in their psychological characteristics. It is their ability to promote growth and their determination to outperform rivals that drives the company's creative production (Marzi & Caputo, 2019; Pavitt, 1998).

In this definition, the entrepreneur can be seen at any stage within an organisation, so it is important to distinguish between capital contributors and entrepreneurial figures (Ciappei, 2005). It is recognisable from this newly defined perspective that the entrepreneur's function does not include taking on risks that are entirely transferred within the property.

From the classic Schumpeterian view, the entrepreneur is the key character in the phase of creating a system and the primary innovator who creates "fresh mixes" in the value chain. Market dominance and wealth are generated through innovation, but the mechanism must be disrupted for this to work; innovation causes profit imbalance (Pavitt, 1998). This is how Schumpeter defined "creative destruction".

Schumpeter (1934) portrays the "heroic" entrepreneur as a pure innovator in a romanticised version of it. The organisation changes in response to the entrepreneur's disruptive powers, introducing a segment of the "chain" of value to each dramatic transition. In this accurate, albeit admirable, philosophy, we see the prodromes of what will inevitably become the grand strand of the value creation theory (Marzi & Caputo, 2019; Pavitt, 1998).

In contrast to the Schumpeterian thought, Knight (1921) based his hypothesis on uncertainty. He defined it as the information available to operators, which is almost always insufficient, resulting in profit opportunities for those who have information about the market (Brouwer, 2002). By doing business in such ambiguous circumstances, the entrepreneur, according to this viewpoint, must be a skilled governor of the state of ambiguity. As a result, someone would be making decisions in an unpredictable and uninsurable environment, but with the possibility of "entrepreneurial gains" (Wales et al., 2013). In this case, the entrepreneurial role does not imply control of the means of production, as it did for Schumpeter (Long, 1983).

In the matter of leadership, the entrepreneur must provide purpose for the employees to motivate workers to achieve goals and adhere to principles (Ciappei, 2005). In contrast to the manager and especially the figure specified by science management, the entrepreneur shares new concepts and ideals, later recombining and consolidating these concepts and ideals into an adhesive that holds all assets of the company together: a genuine leader inspires others to perform the extraordinary (Marzi & Caputo, 2019; Pellegrini et al., 2021).

Consequently, top management must comprehend and address the organisation's problem, thereby initiating a phase of traditional organisational renewal through the decentralisation of activities across the entire system. The entrepreneur retains only the political, operational and harmonisation roles with internal and external partners (Ciappei, 2005). This role is obtained by performing these activities de facto, and delegating them will result in de-entrepreneurialisation. It is apparent that the management of the company can be geared towards a long-term strategy that incorporates the Schumpeterian concept of constant innovation.

The Austrian school advocates an idea that is more concerned with the position of knowledge, and its most prominent proponent is Israel Meir Kirzner (1978). Kirzner contends that competition and entrepreneurship are mutually beneficial and that the entrepreneurial player is central to the operation of the business. The author also draws parallels with Schumpeter's idea, who proposed that the market is a process through which the entrepreneur can identify and exploit untapped opportunities by working within the flow. According to this viewpoint, the entrepreneur serves as the system's driver, encouraging cooperation, trade and opportunity utilisation. Their strategy is to be as perceptive as possible to detect weak signals and seize opportunities ahead of the competition (Hébert & Link, 1989). The entrepreneur is more than just a follower of a pre-determined strategy; they are also a quick scout who can spot opportunities that others would miss (Kirzner, 1978). As a result, certain behaviours must be classified as entrepreneurial decisions focusing on the future, where means are selected and preferences are ordered on the basis of an economisation criterion.

Afterwards, Kirzner (1978) tried to differentiate pure enterprise from the rest of the corporate environment. The author claims that because a pure entrepreneur lacks output capital, this abstraction can be used to define pure entrepreneurial variables (Hébert & Link, 1989).

The author frequently addresses another critical issue that must be elucidated: ambiguity. Ambiguity is linked to two other concepts in Kirzner's theory: time and arbitrage. In the long term, uncertainty is essential, whereas in the short term it is not. From this perspective, arbitrage is the strongest definition of pure entrepreneurship, and Kirzner does not differentiate between the two because he believes that they are related (Sundqvist et al., 2012).

According to Kirzner, arbitrage requires no financial resources and consists of recognising only a profit advantage (in this instance, the difference in price), taking it while purchasing and selling, simultaneously. Therefore, arbitrage is the most appropriate and the most fundamental entrepreneurial practice. When a person requires financial capital but sees a potential for profit that is yet to be realised, they will buy and sell an asset, simultaneously (Sundqvist et al., 2012).

To summarise, Kirzner made a noteworthy contribution to the theory of entrepreneurship because of the high emphasis he placed on this role, analysing it in depth and debunking some conventional economy theories about entrepreneurship, such as the entrepreneur being merely a selector of capital to be introduced into a black-box business (Hébert & Link, 1989). The ability to discover and exploit market opportunities is central to Kirzner's thought; while opportunities remain subjective, the focal point of the matter is this: a specific observed object only transforms into an opportunity when it is mastered and exploited.

2.2   Generating Value from Innovation in an Uncertain World

We may say that the primary goal of an entrepreneur and the growth of a business is the creation of value (Sassetti et al., 2018). Entrepreneurship is described by a desire to act aimed at capturing and capitalising on resources, a project aimed at "doing" (Marzi & Caputo, 2019). The entrepreneur is an individual in action who chooses ideas but, most importantly, acts on them—being timely, and often opportunistic, in leveraging the opportunities that present themselves.

The entrepreneur can be a planner, an innovator, a visionary or an individual who makes decisions and takes risks, specifically, showing that the entrepreneurial vision cannot be contained within the confines of the business but embodied in a wide range of fields (Ciappei, 2005; Pavitt, 1998; Stewart Jr & Roth, 2001).

Nevertheless, pure effort does not become entrepreneurship until it is linked to a conflict for value production; the entrepreneurial mindset is still the driver of the enterprise, yet the proceeds and the management mechanism have to be beneficial for the realisation of those decisions (Marzi & Caputo, 2019; Stewart Jr & Roth, 2001).

Then, it is assumed that the company is managed by people and it is embedded in a dynamic market to generate profit that is satisfying to shareholders and stakeholders. This goal can be attained if competition is overcome by creating competitive advantages, investing and making value-creating decisions.

The basic principle is that a business must place itself optimally in the world and utilise its capabilities and abilities to produce performance that exceeds the average. Therefore, an entrepreneurial attitude that aims at changing the actual situation by interacting with the competitive environment and consumer preferences must have tension as a core component (Hechavarria et al., 2017).

Hence, profit development occurs in the needs–resources partnership or where an organisation realises that the product-market mix will fill a need. As a result, the entrepreneurial tension would be directed at exploiting the gap between the importance of the client's desire and the service value that the organisation requires to meet the said need. The problem of value production is particularly relevant in the distance that can be generated by the organisation: the larger the gap, the greater the benefit (Ciappei, 2005; Hechavarria et al., 2017).

In contrast to the entrepreneur, the manager's decision-making authority is more constrained. Actually, management refers to the degree of professionalism exhibited by the people who are coping with and managing in compliance with the techniques that are outlined by the entrepreneurial top administration. It is linked to the willingness of a manager to accurately observe the process steps, allowing for responsibly made decisions and risk awareness. The strictly relevant topics in this field are mostly related to strategy, coordination and regulation (Caputo & Pellegrini, 2019).

Entrepreneurship and management abilities are equally essential for a company's long-term success: such practice necessitates imagination, intuition and effort, as well as organisation and planning (Kraus & Kauranen, 2009; Marzi & Caputo, 2019; Pellegrini et al., 2021).

In general, management is scarce in freshly formed firms as it becomes overly present in centralised realities; in comparison, a rediscovery of creativity is necessary in certain realities that already lead in the direction of consolidation and work bureaucratisation, leaving little space for inner intervention (Kraus & Kauranen, 2009).

Creativity is a distinguishing characteristic of human behaviour. Creativity is ruled by an intellect that is not logical and is more pronounced in certain people who have the ability to generate uniqueness and transform reality as a result of their capacity to sense new associations between thoughts and things (Edwards-Schachter et al., 2015).

Creativity is therefore something very specific and not a fluid that pervades all directions; a person can have creativity and originality in one way, but not so much in another.

Creative individuals are those who are capable of inventing something unique that will be valued in a specific area (Shu et al., 2020). Creativity occurs when a variety of critical factors come together, including experience, the capacity for creative thinking, enthusiasm (internal motivation) and constancy (that comes from passion) as shown by Sassetti (2021) and Sassetti et al. (2018). Creativity cannot be paralleled with nonconformity and instincts at all costs but rather as the product of genuine mental exercises (Groth & Peters, 1999).

Although the words "creativity" and "innovation" are often used interchangeably, there are certain fundamental distinctions between the two terms. Creativity transcends its position as a fundamental component of invention and can be described as the application of innovative concepts.

If execution is the process of transforming a concept into reality, innovation occurs concurrently with the emergence of the idea. Creativity is a necessary component of innovation; it serves as a springboard (Cropley et al., 2011). Creativity is also described as the creation of ideas, while invention is the process by which these ideas are transformed into behaviour by selection, change and execution (Edwards-Schachter et al., 2015).

The widely held idea that creativity and innovation are inextricably linked is informed by the belief that internal processes in the company are dependent on external processes. Industrial sectors are often characterised by continuous changes as an outcome of technical changes in a so-called technology innovation chain. In other terms, creativity will be turned into invention; creativity will serve as the input, and innovation will serve as the output (Groth & Peters, 1999).

Thus, creativity is essentially the act of generating ideas, and innovation is the product of a filtering process that seeks the separation of good and bad ideas from each other (Edwards-Schachter et al., 2015; Marzi & Caputo, 2019).

Scholars postulate that innovation should be considered from a process viewpoint, identifying a direction forward through pathways of invention and feasibility tests to achieve commercialisation and organisational changes (McAdam & McClelland, 2002; Smith et al., 2008).

The first stage in innovation therefore entails concept generation using many of the methods that are now common practice in the business environment, such as design thinking or "interpretive engagement" (Brown, 2008; Thoring & Müller, 2011).

The second step will determine compliance with the company's goals and ensure that they are consistent and aligned with the purpose and goals of the organisation, as well as with the imposed financial restrictions.

The third step is an examination of the technological and economic viability of the new project, as well as the existence of required competencies—material and immaterial—and its realisation. In the end, the execution refers to the mechanism of converting a concept into an action and the entrepreneur is the pinnacle of this process component throughout all the methods of optimisation and formalisation (McAdam & McClelland, 2002).

The literature on entrepreneurship also outlines some essential characteristics of the entrepreneur as an innovator (Hood & Young, 1993). Overall, these are the desire to prepare, the activism towards action, and the promptness in capitalising on opportunities, which are complemented by a set of critical entrepreneurial characteristics, including innovation, promptness, taking risks, leadership and friction in value production (Hood & Young, 1993; Marzi & Caputo, 2019).

We have already addressed Schumpeter's, Kirzner's and Knight's contributions to the first three, although the idea of leadership merits more enquiry.

In this regard, it is crucial to note that the analysis of this specific function began with individuals who seemed to be outside the corporate community. This is the stance of sociologist Max Weber (Weber, 2002), who conceived the idea of an inspirational leader as an exceptional individual with an atmosphere of sacredness and heroism, a figure who sets the standard and guides other people (Brouwer, 2002). In this situation, it is the economic performance that causes people to obey and trust a person, as the "evidence" is what distinguishes the person as a chief. Furthermore, charisma is characterised by Weber (2002) as a transformative force that generates the past and brings into existence a shift of precepts that affects the culture of organisation. Weber's (2002) inspirational quality seems to be consistent with Schumpeter's ideas about change that ruins the past (Brouwer, 2002).

Another important contribution to this trend is that of Tichy and Devanna (1986), who created a model of transformational leadership based on a deeply ambiguous and continuously shifting antecedent. According to the writers, global challenges and changing business dynamics are the root causes of the company's transition phase. The role of the business captain is therefore to lead the enterprise through the transition phase while attempting to ease the conflicts caused by this challenge.

2.3   The Role of Knowledge in Innovation

Creating value begins with the development and accumulation of knowledge, which is at the heart of innovation (Thornhill, 2006). Indeed, the value of expertise, knowledge and learning in companies and workplaces has grown since the 1980s (Pellegrini et al., 2020, 2021).

It is generally recognised that the accumulation and collection of knowledge possessed by individuals within an enterprise will result in a highly competitive advantage. The growth of internal entrepreneurship and the sphere of knowledge are subjects that the literature has gradually deemed increasingly significant compared to the existing model of organisation and management (Thornhill, 2006).

It is now apparent that continuous innovation is a fundamental and necessary condition for staying competitive in volatile and unceasingly changing markets. Businesses are driven by volatility to pursue expertise that is beyond the organisation's bulwarks, seeking not only a collection of technological skills but also the capacity for human development and metacognitive abilities that enable renewal and constant adaptation to an evolving external environment (Corso et al., 2003).

It is worthwhile to explain the definition of knowledge's roots, which can be traced back to classical Greece and, more precisely, to Plato's insights, who described knowledge as a belief that has been shown to be real (White, 1976). According to Plato's philosophy, the real world and our understanding of it represent a different world to which the person aspires but cannot directly recognise, as reflected in the famous cave legend.

Regarding management studies on knowledge, Alfred Marshall (2009) is a name that must be mentioned since he was the first to recognise that the enterprise's resources come from the knowledge the organisation creates. Later, Nelson and Winter (1977) made another significant contribution, within their economic and technical transition evolutionary theory that a knowledge record is preserved in repetitive daily activity trends and thus in routines. It is because of these writers that the idea of repetitive actions within an enterprise found its way into economic organisational philosophy, thus opening the field to much of the foundational literature on technical orientations that first linked innovation with knowledge (Popadiuk & Choo, 2006).

Many scholars then pondered how knowledge may be applied in management through the convergence of managerial approaches and organisational levels. Chester I. Barnard (1940) was one of those who argued that it had a behavioural and unconscious dimension, as well as a logical-linguistic part.

Herbert Simon (2013), citing Bernard's intuitions, also finds knowledge to be a problem-solving capability as well as a cybernetic information retrieval approach. Human beings, according to the author's model, function as mechanisms that process information, extrapolating sensory inputs from purposeful constructs derived from their experience and then preserving these experiences as new knowledge.

Simon's thinking was undoubtedly inspired by the first moves made by psychology towards cybernetics. The author's importance, though, is in emphasising the rational aspect of human thinking and the processes of making organisational decisions, as well as the limitations of cognitive capacities that humans have, which are now commonly studied as cognitive biases.

Following that, Cohen et al. (1972) questioned Simon's paradigm by highlighting the features of complexity and irrational methods for making decisions and solving problems. For the writers, the organisation and the people who work for it exist in a world full of questions and expectations, where resources are the garbage and difficulties are the can (see the "garbage can" model). The three writers of this study emphasise that deductive strategy choices have significant shortcomings and are only significant, retrospectively. As a result, it is assumed that learning occurs only on a personal and non-organisational basis and that it is impossible to produce mutual and diffused knowledge (Nonaka, 1994).

From evolutionary viewpoints, organisations and cultures are prone to constant alterations and transformation, and after World War II, a growing majority of developed countries transitioned from a system of industrial economy to a service economy, eventually winding up in what we now know as information and knowledge economics.

Peter Drucker (2007), who invented the words "knowledge job" and "knowledge worker" in the 1960s, is one scholar who succeeded in understanding the significance of this transition, stressing how important knowledge is as an economic resource through which organisations could face and develop ways to navigate their processes of learning and transformation. Specific talents, according to the author, cannot be converted into writing or published and therefore may only be mastered by gaining experience, being tacit knowledge within the organisation. Organisations adapt, develop and enhance their routines by gaining the knowledge of their individuals and starting an internal learning mechanism that shows how the growth of the person leads to the transformation of the world in which it works (Drucker & Zahra, 2003).

Senge (1994) acknowledged that certain organisations are deficient and they do not encourage the members of their organisations to acquire the wealth of knowledge held within them or to build new ones in the best possible ways. Senge further criticises the reductionist mentality prevalent in Western thinking, which seeks to oversimplify issues with a more nuanced nature, resulting in a lack of variety in the foundational knowledge for corporate development and learning (Kofman & Senge, 1993).

The latest contributions were made in the 1990s, with the method proposed by Prahalad and Hamel (1994), which centred on distinct knowledge, and the proposal by Stalk et al. (1992), which concentrated on skill-based rivalry.

This community of scholars have stressed how, in the 1990s, the industry underwent significant transformations that rendered the hierarchical strategy redundant, as a company's dynamic capacity is described as the ability to plan, understand, adjust and constantly change to the external background, at strategic and operational levels. As a result, the behavioural and immaterial value of certain tools is stressed, coupled with a complex capacity for regeneration that allows for the continual improvement of the main economic structures in the company, referred to as strategic capacities to gain sustainable effectiveness and dominance over rivals.

According to this perspective, it is critical that the required elements should attain a more competitive advantage in the inner resources and skills than in the outer environment. Consequently, it is evident that creativity and knowledge, more tacitly expressed, hold a critical part in setting higher performance. It is the undisclosed knowledge that allows a company to be different from competitors while using innovation as a competitive leverage (Nonaka, 1994).

REFERENCES

Barnard, C. (1940). Comments on the job of the executive. *Harvard Business Review, 18*, 295–308.

Brouwer, M. T. (2002). Weber, Schumpeter and Knight on entrepreneurship and economic development. *Journal of Evolutionary Economics, 12*(1), 83–105.

Brown, T. (2008). Design thinking. *Harvard Business Review, 86*(6), 84.

Caputo, A., & Pellegrini, M. M. (Eds.). (2019). *The anatomy of entrepreneurial decisions: Past, present and future research directions.* Springer.

Ciappei, C. (2005). *Strategia e valore di impresa.* Firenze University Press.

Cohen, M. D., March, J. G., & Olsen, J. P. (1972). A garbage can model of organizational choice. *Administrative Science Quarterly, 17*(1).

Corso, M., Martini, A., Pellegrini, L., & Paolucci, E. (2003). Technological and organizational tools for knowledge management: In search of configurations. *Small Business Economics, 21*(4), 397–408.

Cropley, D. H., Kaufman, J. C., & Cropley, A. J. (2011). Measuring creativity for innovation management. *Journal of Technology Management & Innovation, 6*(3), 13–30.

Drucker, P., & Zahra, S. A. (2003). An interview with Peter Drucker. *The Academy of Management Executive (1993–2005)*, 9–12.

Drucker, P. F. (2007). *Innovation and entrepreneurship: Practice and principles.* Routledge.

Edwards-Schachter, M., García-Granero, A., Sánchez-Barrioluengo, M., Quesada-Pineda, H., & Amara, N. (2015). Disentangling competences: Interrelationships on creativity, innovation and entrepreneurship. *Thinking Skills and Creativity, 16*, 27–39.

Groth, J., & Peters, J. (1999). What blocks creativity? A managerial perspective. *Creativity and Innovation Management, 8*(3), 179–187.

Hébert, R. F., & Link, A. N. (1989). In search of the meaning of entrepreneurship. *Small Business Economics, 1*(1), 39–49.

Hechavarria, D. M., Terjesen, S. A., Ingram, A. E., Renko, M., Justo, R., & Elam, A. (2017). Taking care of business: The impact of culture and gender on entrepreneurs' blended value creation goals. *Small Business Economics, 48*(1), 225–257.

Hood, J. N., & Young, J. E. (1993). Entrepreneurship's requisite areas of development: A survey of top executives in successful entrepreneurial firms. *Journal of Business Venturing, 8*(2), 115–135.

Kirzner, I. M. (1978). *Competition and entrepreneurship.* University of Chicago Press.

Knight, F. H. (1921). *Risk, uncertainty, and profit.* Hart, Schaffner & Marx; Houghton Mifflin Co.

Kofman, F., & Senge, P. M. (1993). Communities of commitment: The heart of learning organizations. *Organizational Dynamics, 22*(2), 5–23.

Kraus, S., & Kauranen, I. (2009). Strategic management and entrepreneurship: Friends or foes? *International Journal of Business Science & Applied Management (IJBSAM), 4*(1), 37–50.

Long, W. (1983). The meaning of entrepreneurship. *American Journal of small business, 8*(2), 47–59.

Marshall, A. (2009). *Principles of economics: Unabridged* (8th ed.). Cosimo.

Marzi, G., & Caputo, A. (2019). *Responsible entrepreneurship education: Emerging research and opportunities.* IGI Global.

McAdam, R., & McClelland, J. (2002). Individual and team-based idea generation within innovation management: Organisational and research agendas. *European Journal of Innovation Management.*

Nelson, R. R., & Winter, S. G. (1977). In search of useful theory of innovation. *Research Policy, 6*(1), 36–76.

Nonaka, I. (1994). A dynamic theory of organizational knowledge creation. *Organization Science, 5*(1), 14–37.

Pavitt, K. (1998). Technologies, products and organization in the innovating firm: What Adam Smith tells us and Joseph Schumpeter doesn't. *Industrial and Corporate Change, 7*(3), 433–452.

Pellegrini, M. M., Ciampi, F., Marzi, G., & Orlando, B. (2020). The relationship between knowledge management and leadership: Mapping the field and providing future research avenues. *Journal of Knowledge Management.*

Pellegrini, M. M., Ciappei, C., Marzi, G., Dabić, M., & Egri, C. P. (2021). A philosophical approach to entrepreneurship education: A model based on Kantian and Aristotelian thought. *International Journal of Entrepreneurship and Small Business, 42*(1–2), 203–231.

Popadiuk, S., & Choo, C. W. (2006). Innovation and knowledge creation: How are these concepts related? *International Journal of Information Management, 26*(4), 302–312.

Prahalad, C. K., & Hamel, G. (1994). Strategy as a field of study: Why search for a new paradigm? *Strategic Management Journal, 15*(S2), 5–16.

Sassetti, S. (2021). *Entrepreneurship and emotions: Insights on venture performance.* Emerald Group Publishing.

Sassetti, S., Marzi, G., Cavaliere, V., & Ciappei, C. (2018). Entrepreneurial cognition and socially situated approach: A systematic and bibliometric analysis. *Scientometrics, 116*(3), 1675–1718.

Schumpeter, J. A. (1934). *Theory of economic development.* Harvard University Press.

Senge, P. M. (Ed.). (1994). *The fifth discipline fieldbook: Strategies and tools for building a learning organization.* Random House LLC.

Shu, Y., Ho, S. J., & Huang, T. C. (2020). The development of a sustainability-oriented creativity, innovation, and entrepreneurship education framework: A perspective study. *Frontiers in Psychology, 11*, 1878.

Simon, H. A. (2013). *Administrative behavior.* Simon and Schuster.

Smith, M., Busi, M., Ball, P., & Van der Meer, R. (2008). Factors influencing an organisation's ability to manage innovation: A structured literature review and conceptual model. *International Journal of Innovation Management, 12*(04), 655–676.

Stalk, G., Evans, P., & Sgulman, L. E. (1992). *Competing on capabilities: The new rules of corporate strategy* (Vol. 63). Harvard Business Review.

Stewart, W. H., Jr., & Roth, P. L. (2001). Risk propensity differences between entrepreneurs and managers: A meta-analytic review. *Journal of Applied Psychology, 86*(1), 145.

Sundqvist, S., Kyläheiko, K., Kuivalainen, O., & Cadogan, J. W. (2012). Kirznerian and Schumpeterian entrepreneurial-oriented behavior in turbulent export markets. *International Marketing Review.*

Thoring, K., & Müller, R. M. (2011, September 8–9). Understanding design thinking: A process model based on method engineering. In *DS 69: Proceedings of E&PDE 2011, the 13th International Conference on Engineering and Product Design Education* (pp. 493–498).

Thornhill, S. (2006). Knowledge, innovation and firm performance in high-and low-technology regimes. *Journal of Business Venturing, 21*(5), 687–703.

Tichy, N. M., & Devanna, M. A. (1986). The transformational leader. *Training & Development Journal.*

Wales, W. J., Parida, V., & Patel, P. C. (2013). Too much of a good thing? Absorptive capacity, firm performance, and the moderating role of entrepreneurial orientation. *Strategic Management Journal, 34*(5), 622–633.

Weber, M. (2002). *The Protestant ethic and the Spirit of capitalism: And other writings.* Penguin.

White, N. P. (1976). *Plato on knowledge and reality.* Hackett Publishing.

CHAPTER 3

The Necessity and the Drawbacks of Product and Process Innovation

Abstract Based on data from the Scopus database spanning from 1990 to 2021, the present chapter aims to provide a summary of articles, research streams and the most significant publications addressing the role of product and process innovation in a manufacturing context. Following a short survey of the most important studies on this subject, the current chapter discusses the main journals, the most relevant keywords and the primary research streams in this area of study by means of visualisation of similarities keyword analysis. The findings show that developing new products and improving production processes have a positive impact on firms' competitiveness. However, the literature also shows that some drawbacks exist, requiring careful consideration by entrepreneurs about adopting or not adopting a specific type of innovation. A case of innovation in manufacturing firms is presented, highlighting the double nature of product and process innovation as well as the beneficial and harmful effects of adopting certain types of innovations.

Keywords Product • Process • Innovation • Manufacturing • VOS

© The Author(s), under exclusive license to Springer Nature   21
Switzerland AG 2022
G. Marzi, *Uncertainty-driven Innovation*,
https://doi.org/10.1007/978-3-030-99534-8_3

3.1   Different Types of Innovation

The innovation process is critical for performance improvement, competitive advantage sustainability and firm survival (Damanpour, 1991; Smith & Tushman, 2005). By meeting the demands of relevant consumers and implementing new products and processes, companies are enabled to penetrate or build new markets (Marzi et al., 2017).

For example, Lukas and Ferrell (2000) argue that adapting the launch of new products to consumer needs is a crucial condition for maintaining a competitive position in increasingly technologically advanced markets. In this regard, the process by which a business gains additional information and expertise is crucial for the company's establishment of innovative products or services (Knudsen & Levinthal, 2007).

The process of innovation is based on the usage, recognition and utilisation of innovations and process improvements by both managers and entrepreneurs (Marzi & Caputo, 2019). Thus, it is evident that innovation has a major impact on the internal behaviours and relationships of organisations as well as on the strategies and processes of the business (Marzi et al., 2018).

The written works on organisational theories of innovation have concentrated on identifying possible classifications for this phenomenon, including the classic distinctions between (a) administrative or technical innovation in relation to the organisational process at hand and (b) product or process innovation in relation to the particular innovation object (Nord & Tucker, 1987).

The second difference between product and process innovation is deemed significant for pursuing a sustainable advantage in competition. In the manufacturing context, product innovations refer to new products and services brought to the market, generally to satisfy latent customer needs (Damanpour, 1991; Lukas & Ferrell, 2000). Process innovation refers to additional elements embedded into the operation and development processes of the industry, such as new materials and machinery for the manufacturing company (Damanpour, 1991).

Innovation in manufacturing firms is distinct from innovation in companies providing services (Becheikh et al., 2006) and many investments in literature concentrate on the study of manufacturing innovation (Becheikh et al., 2006; Terziovski, 2010).

For example, Sirilli and Evangelista (1998) discovered that when comparing the features of innovation processes in manufacturing and service companies, process innovations are more prevalent in service firms, while product innovation is considered more significant in the majority of manufacturing firms.

This is also shown in a study of business managers conducted by Linder et al. (2003), which shows that business decision makers place a higher premium on new products than on process improvements. According to Sirilli and Evangelista (1998), innovation costs nearly three times as much in the manufacturing sector as it does in the service sector. According to Becheikh et al.'s (2006) research on innovation in manufacturing companies, a sizeable portion of the literature focuses solely on product innovation, while just a few writings are concerned with process innovation. Because scholars have lower interest in process innovation, it is often considered less significant than product innovation.

However, some studies show that the two forms of innovation are inextricably related and not mutually exclusive. As a result, ignoring process innovations could impair a firm's capability to introduce product innovations, thus harming the whole process of innovation (Becheikh et al., 2006).

Process innovation boosts businesses' productivity, resulting in competitive advantages primarily through cost reductions and increased versatility of the manufacturing. Process innovation also has the ability to promote product innovation (Hall et al., 2009; Martinez-Ros, 1999).

Given that innovation promotes firm growth, internationalisation and performance, the latest writings have also extensively put innovativeness at the centre of attention in small and medium-sized enterprises (SMEs), alleviating the extensive absence of interest in the phenomenon of innovation in such companies (Marzi et al., 2017, 2018). Improving the innovativeness of SMEs appears to be critical for communities and regions to develop economically, fostering strategic alliances and collaborations among such firms.

Laursen and Salter (2006) discovered that SMEs exhibit the same levels of innovation as large corporations, particularly in terms of radical innovation. Interestingly, the body of research also emphasised the importance of open innovation for SMEs to develop, arguing that whereas the focal point in large companies primarily is on Research and Development (R&D) in open innovation efforts, the focal point of SMEs is increased on commercialisation because, even though a large number of SMEs hold a good positioning in terms of innovation capabilities, they frequently lack the capacity when it comes to facilities for manufacturing, global contacts to put forward new products and marketing channels.

In this vein, Van de Vrande et al. (2009) examined the impact of open innovation on SMEs and discovered that small businesses are increasingly playing a visible role in situations of modern innovation. In particular, the authors emphasise how innovation in SMEs is hindered by a shortage of financial resources, scarce chances to take on specialised workers and small innovation portfolios, which means it is not possible to overcome the dangers associated with innovation.

To find missing innovation resources, SMEs must be heavily dependent on the networks they have. In accordance with these findings, Chang (2011) offered a few hypotheses on the topic, stating that knowledge development and improvement increase the firm's exploratory and exploitative innovative ability. A significant level of dynamism and competitiveness has a positive correlation with innovation in SMEs.

It is clear that innovation is critical for the organisational, technological and strategic development of SMEs. Evangelista et al. (1997) demonstrate the presence of two distinct models of innovation in manufacturing firms, namely the model of large companies dependent on R&D investments and the model of SME innovation characterised by innovation activities of an informal nature.

SMEs often differ from large companies in terms of their spending on innovation. Although large companies spend heavily on research and development, the primary expenditure in SMEs is on the purchase of new equipment, machinery and facilities to promote innovation (Evangelista et al., 1997). Innovation is important for SMEs to address the disadvantages associated with operating in a globalised world (Marzi et al., 2018). To build value in such a globalised world, it is important to keep innovating and using new chances continuously in an effort to maintain a sustainable competitive advantage (Hurmelinna-Laukkanen et al., 2008).

However, because of the unique characteristics of SMEs, managers must be alert to the limits and risks associated with implementing innovation in the organisational structures of the SMEs. As a result, investments based on knowledge, strategic alliances, and the awareness and motivations of entrepreneurs regarding this phenomenon are critical components to consider.

3.2 DATA GATHERING AND ANALYSIS

The extensive literature exploration on product and process innovation in manufacturing companies started with a collection of published articles on the topic. The data gathering and analysis procedure have been described in detail below for each step, allowing a full evaluation and replication by other researchers (Oldendick, 2012).

For the purpose of this study, Scopus was selected because it has a collection of articles that would have important and profound effects on this study. The articles in Scopus have also surpassed stringent criteria for impact factor and citation count.

The collection of data was achieved via a research query using the subsequent research words limited to the English language, "Article" as text forms and a time period of 1990–2021:

TITLE-ABS-KEY (manufactur*) AND TITLE-ABS-KEY ("product innovation" OR "process innovation")

On the Advanced Research page, "TITLE-ABS-KEY" stands for "Title, Abstract and Keywords". It retrieved 1791 entries from all study areas. Then the sample search was narrowed using "Business, Management and Accounting" as search term for the research area. A total dataset of 1116 articles and reviews was obtained.

The final query on Scopus resulted as follows:

TITLE-ABS-KEY (manufactur*) AND TITLE-ABS-KEY("product innovation" OR "process innovation") AND (LIMIT-TO (DOCTYPE, "ar")OR LIMIT-TO (DOCTYPE, "re")) AND (LIMIT-TO (PUBYEAR, 2021) OR LIMIT-TO (PUBYEAR, 2020) OR LIMIT-TO (PUBYEAR, 2019) OR LIMIT-TO (PUBYEAR, 2018) OR LIMIT-TO (PUBYEAR, 2017) OR LIMIT-TO (PUBYEAR, 2016) OR LIMIT-TO (PUBYEAR, 2015) OR LIMIT-TO (PUBYEAR, 2014) OR LIMIT-TO (PUBYEAR, 2013) OR LIMIT-TO (PUBYEAR, 2012) OR LIMIT-TO (PUBYEAR, 2011) OR LIMIT-TO (PUBYEAR, 2010) OR LIMIT-TO (PUBYEAR, 2009) OR LIMIT-TO (PUBYEAR, 2008) OR LIMIT-TO (PUBYEAR, 2007) OR LIMIT-TO (PUBYEAR, 2006) OR LIMIT-TO (PUBYEAR, 2005) OR LIMIT-TO (PUBYEAR, 2004) OR LIMIT-TO (PUBYEAR, 2003) OR LIMIT-TO (PUBYEAR, 2002) OR LIMIT-TO (PUBYEAR, 2001) OR LIMIT-TO (PUBYEAR, 2000) OR LIMIT-TO (PUBYEAR, 1999) OR LIMIT-TO (PUBYEAR, 1998) OR LIMIT-TO (PUBYEAR, 1997) OR LIMIT-TO (PUBYEAR, 1996) OR LIMIT-TO (PUBYEAR, 1995) OR LIMIT-TO (PUBYEAR, 1994) OR LIMIT-TO (PUBYEAR,

1993) OR LIMIT-TO (PUBYEAR, 1992) OR LIMIT-TO (PUBYEAR, 1991) OR LIMIT-TO (PUBYEAR, 1990)) AND (LIMIT-TO (SUBJAREA, "BUSI"))

The dataset therefore included journals that are competitively ranked among the most widely cited and influential core journals in the business area (Leydesdorff et al., 2013).

Following a short presentation of articles, authors and journals in this area of science, VOSviewer 1.6.17 was used for the text-mining routine to chart the research streams using the many keywords in the papers (Van Eck & Waltman, 2010).

The text-mining routine's map is a plot where the difference between keywords may be viewed as a representation of the terms' relatedness. The closer the gap between keywords, the closer the keywords are to one another. For determination of the relatedness of keywords, co-occurrences in the texts are used (Van Eck et al., 2010).

Following the keyword study, the subsequent action was the analysis of clusters, specifically the measurement of intra- and inter-cluster variety to further understand the knowledge base variety within individual clusters (Van Eck & Waltman, 2014).

The analysis of the clusters demonstrates the richness based on the scholars' knowledge of the topic. When keywords belonged to the same cluster, they formed a strong group. It showed that a cluster reflects a line of study or a specific subject based on similarities. However, in cluster visualisation, proximity must be considered. There were instances where an object belonged to one cluster but was close to another. In such instances, even if the elements did not belong to the same cluster, these elements can have a reasonable relationship to one another, although without the intensity required to be a part of the same cluster. Such a problem arose as a result of the inability to display the visualisation of similarities (VOS) output in the third dimension. The size of a point reflects its number of occurrences, the larger it is, the more occurring it is (Waltman et al., 2010).

In terms of the keyword review, keywords from authors that appeared (occurred) at least 5 times in the dataset have been used, setting a resolution of 1.00 and a minimum cluster size of 15 (Waltman et al., 2010). In total, 112 keywords were extracted and organised into 4 different clusters.

3.3   A BIBLIOMETRIC ANALYSIS OF PRODUCT AND PROCESS INNOVATION IN THE MANUFACTURING FIELD

As shown by the following figure (Fig. 3.1), the topic of product and process innovation in manufacturing companies dramatically evolved over the time. The number of papers on the topic slowly grew annually, reaching a peak of **96** published papers by **2021**. Such an evolution remarks the increasing importance of product and process innovation, especially for the manufacturing companies.

Table 3.1 presents the most prolific journals on the study topic. Notably, *Journal of Product Innovation Management* is the cornerstone journal within this field of study with a remarkable number of 36 papers published over the target period 1990–2021. Next were *Research policy* and *Technovation*. Interestingly, in terms of the number of published papers, *Business Strategy and the Environment* and *Journal of Cleaner Production were* in the fourth and sixth positions, respectively. Such evidence stresses the prominent role of sustainability issues, green product innovation and environmental impact which have gained increasing attention in the innovation management field of study in recent years. Such finding has also

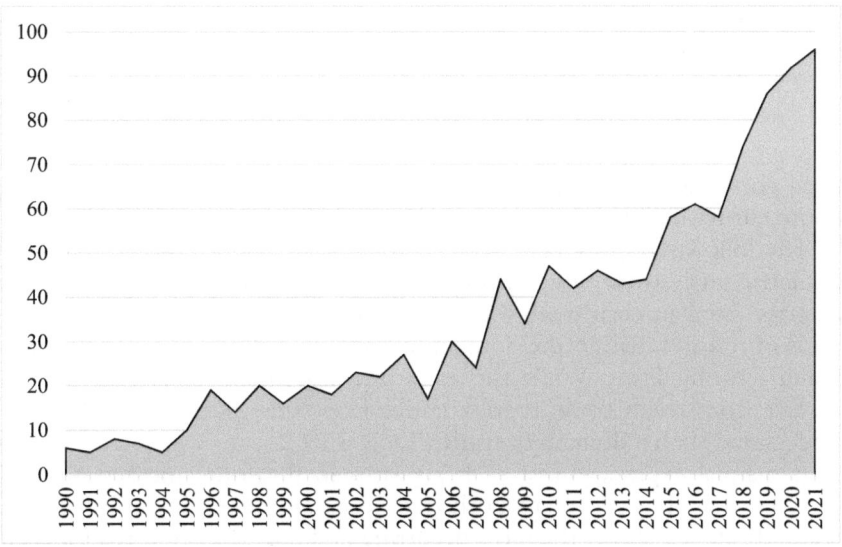

Fig. 3.1   Papers published annually between 1990 and 2021

Table 3.1 Most prolific journals over the years 1990–2021

Journal title	Number of papers
Journal of Product Innovation Management	36
Research Policy	32
Technovation	29
Business Strategy and the Environment	28
International Journal of Production Economics	24
Journal of Cleaner Production	23
Economics of Innovation and New Technology	21
International Journal of Technology Management	19
Journal of Manufacturing Technology Management	19
European Journal of Innovation Management	18
International Journal of Innovation Management	17
International Journal of Operations and Production Management	16
Journal of Business Research	16
International Journal of Production Research	15
IEEE Transactions on Engineering Management	13
Technology Analysis and Strategic Management	13
Small Business Economics	12
Industrial Management and Data Systems	11
Industrial Marketing Management	11
Journal of Technology Management and Innovation	11
Business Process Management Journal	10
Journal of Business and Industrial Marketing	10
Technological Forecasting and Social Change	10

been confirmed by the following keyword analysis, which highlights an entire cluster devoted to it.

The following figure (Fig. 3.2) represents the core of the present bibliometric analysis. Figure 3.2 depicts the graphical output of the VOS analysis. We can clearly see four interweaving clusters representing four areas of research within the domain of product and process innovation in manufacturing firms. While the areas of research are interlaced, we can extract from them some representative keywords, helping us to better understand such a domain of study (Table 3.2).

A manual exploration of each keyword in the four emerged clusters revealed that they represent both traditional and emerging sub-fields of study. Table 3.2 shows the most occurring and relevant keywords for each which allowed the identification of some interesting topics.

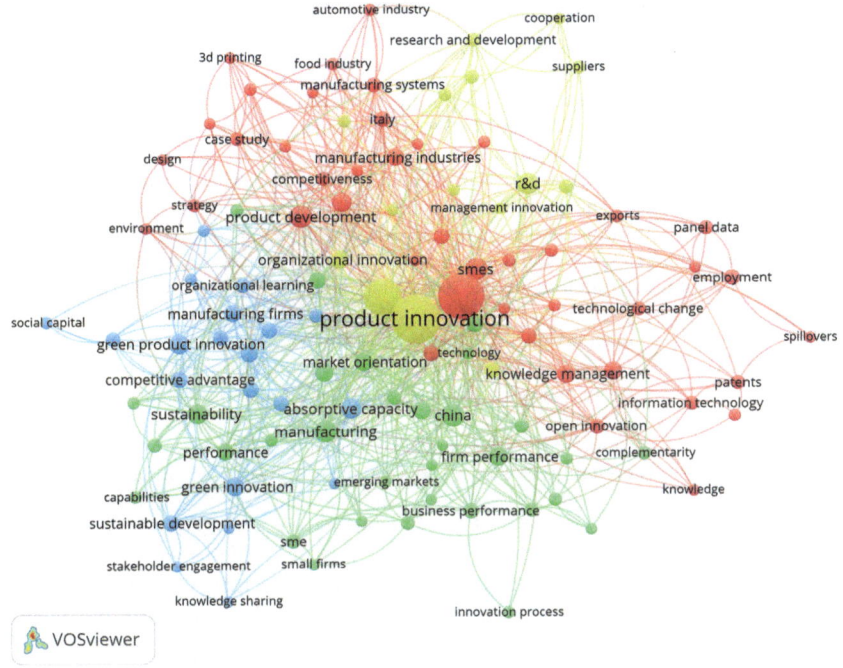

Fig. 3.2   Graphical result of the VOS analysis

The Red cluster was focused on the management of the innovation process, a traditional field of study that includes topics such as "new product development", "knowledge management", "innovation strategy" and "manufacturing systems". However, some interesting themes emerged, such as "industry 4.0", "additive manufacturing" and "3D printing" representing new trends in the innovation topic.

The Green cluster was focused on the role of the entrepreneur and the innovation process. In the yellow cluster, entrepreneurship was explored in relation to traditional entrepreneurship topics such as "market orientation", "business performance", "entrepreneurial orientation" and "intellectual capital". Such relation of keywords stresses the pivotal role of entrepreneurial activity in fostering and guiding the innovation process.

The Blue cluster is the newest and the most rapidly growing area of research and is represented by the relationship of sustainability issues and

Table 3.2   Cluster analysis and representative keywords

Cluster	Representative keywords
Red—The management of the innovation process	New product development; knowledge management; productivity; industry 4.0; innovation strategy; manufacturing systems; open innovation; competitiveness; patents; technological change; additive manufacturing; human capital; new products; 3D printing; automotive industry; design; lean production
Green—The entrepreneur and the innovation process	Entrepreneurship; market orientation; supply chain management; technological innovation; business performance; entrepreneurial orientation; intellectual capital; organisational performance;
Blue—Sustainability issues and innovation	Green product innovation; green innovation; sustainable development; green process innovation; environmental innovation; quality management; environmental regulation; stakeholder engagement
Yellow—R&D management	product innovation; process innovation; R&D; research and development; TQM; cooperation; operational performance

the innovation process, that is, product and process innovation. A sustainable process of innovation cannot be separated from sustainability and environmental issues. As a result, the present cluster is growing, encompassing several facets of sustainability such as "green product innovation", "green process innovation" and "green innovation".

Finally, the yellow cluster represents another traditional fact of product and process innovation for manufacturing companies, the role of R&D management and its relationship with performance and cooperation among different actors.

3.4   An Overview of Influential Studies About Product and Process Innovation in Manufacturing Firms

Product and process innovation in manufacturing firms has been widely investigated over time by different scholars. The body of knowledge about this topic dramatically grew over the years. Table 3.3 illustrates the most influential publications in this field of research showing the topic evolution over the years. The papers have been selected based on the number of citations and the number of normalised citations.

Table 3.3 Most influential publications over the years 1990–2021

Author(s)	Title	Journal	Year	Key Findings
Damanpour, F.	Organizational Size and Innovation	Organization Studies	1992	The article discusses the impact of organisational form on firms and the relationship between firms' small innovative divisions and the benefits that come with being large
Herstatt, C., Von Hippel, E.	From Experience—Developing New Product Concepts via the Lead User Method—A Case Study in a "Low-Tech" Field	Journal of Product Innovation Management	1992	The research shows that the Lead user approach has a range of benefits for product innovation in terms of competition, effort and finance invested, as well as improved end-user satisfaction
Ettlie, J., Reza, E.M.	Organizational Integration and Process Innovation	Academy of Management Journal	1992	The article shows how process innovation can help manufacturing companies achieve a competitive edge and maximise competitiveness through a new organisational system, enhanced manufacturing, design, collaboration, and increased cooperation with suppliers
Capon, N., Farley, J.U., Lehmann, D.R., Hulbert, J.M.	Profiles of Product Innovators Among Large US Manufacturers	Management Science	1992	The article examines the relationship between environment, policy, formal and informal organisation, innovation of products and financial results. It shows that an environment that fosters innovation and collaboration has a beneficial effect on financial gains

(*continued*)

Table 3.3 (continued)

Author(s)	Title	Journal	Year	Key Findings
Banbury, C.M., Mitchell, W.	The Effect of Introducing Important Incremental Innovations on Market Share and Business Survival	Strategic Management Journal	1995	The research shows a close correlation between incremental product innovation and sped up product launch and market success. The paper claims that longevity of a business is primarily dictated by its capability to keep new products in the market, rather than by launching products that are technically innovative
Damanpour, F.	Organizational Complexity and Innovation: Developing and Testing Multiple Contingency Models	Management Science	1996	The study develops a model to better explain the dichotomy between uncertainty and innovation, based on 30 years of data
Atuahene-Gima, K.	Market Orientation and Innovation	Journal of Business Research	1996	The research shows how market focus of a firm and chosen innovation characteristics affect the success of service and product technologies
Atuahene-Gima, K.	Differential Potency of Factors Affecting Innovation Performance in Manufacturing and Services Firms in Australia	Journal of Product Innovation Management	1996	Using Australian companies as a case study, this article examines managers' views of the conditions that contribute to virtuous new product and service creation. The author discovers that production firms prioritise product innovation and efficiency, while service firms prioritise human capital strategies

(*continued*)

Table 3.3 (continued)

Author(s)	Title	Journal	Year	Key Findings
Hatch, Mowery	Process Innovation and Learning by Doing in Semiconductor Manufacturing	Management Science	1998	This paper looks into the relationship between process innovation and active learning. The article shows how gained expertise, with devoted process production facilities and the spatial proximity of development and production equipment, can significantly improve efficiency of implementing new technology
Lukas, B.A., Ferrell, O.C.	The Effect of Market Orientation on Product Innovation	Journal of the Academy of Marketing Science	2000	The research emphasises the significance of consumer orientation to product innovation. The key observation involves consumer focus, which results in an improvement in the launch of innovative products and a decline in introducing me-too products. Additionally, competitor focus results in an improvement in the adoption of similar products and a decrease in introducing line extensions and novel products

(*continued*)

Table 3.3 (continued)

Author(s)	Title	Journal	Year	Key Findings
Mathieu, V.	Service strategies within the manufacturing sector: benefits, costs and partnership	International Journal of Service Industry Management	2001	The article examines the risks and advantages of service strategies in industrial companies. It categorises strategies according to their service specificity and organisational strength, elaborating on the relationship among both elements and the likelihood of running a cooperative choice. It is concluded by the author that while the more ambitious tactics offer greater advantages to companies; they are often the most risky due to the many hazards associated with them
Thomke, S., Von Hippel, E.	Customers as Innovators—A New Way to Create Value	Harvard Business Review	2002	The paper emphasises the importance of paying attention to consumer expectations when developing product innovation trajectories. This is accomplished by including consumers in the process of designing and testing processes. The article discusses the outcomes of this innovative strategy, which include cost savings, increased generated value, and increased customer loyalty

(*continued*)

Table 3.3 (continued)

Author(s)	Title	Journal	Year	Key Findings
Rogers, M.	Networks, Firm Size and Innovation	Small Business Economics	2004	The papers compared manufacturing companies to non-manufacturing companies in terms of innovation and networking. In particular, when it comes to innovation, small companies working in manufacturing tend to have more network related benefits than medium and large size companies
Koufteros, X., Vonderembse, M.A., Jayaram, J.	Internal and external integration for product development: the contingency effect of uncertainty, equivocality, and platform strategy	Decision Sciences	2005	The authors emphasised how a high level of inner integration is connected with a high level of outer integration, as well as the importance of contextual variables in moderating the relationship between integration strategy and success. They establish that both convergence and profitability hold a beneficial effect on product innovation and profitability. Additionally, the findings suggest that ambiguity acts as a moderator in the relationships between integration and efficiency

(*continued*)

Table 3.3 (continued)

Author(s)	Title	Journal	Year	Key Findings
Nieto, M.J., Santamaría, L.	The Importance of Diverse Collaborative Networks for the Novelty of Product Innovation	Technovation	2007	The research emphasises the relevance of partnership by demonstrating that expertise managing partnerships result in an improved product innovation outcomes and that the collaborative network's partner selection can be a critical factor in the progress of innovation
Cooper, R.G.	Perspective: The Stage-Gate® Idea-to-Launch Process—Update, What's New, and NexGen Systems	Journal of Product Innovation Management	2008	The article provides an in-depth study of the advantages of the Stage-Gate approach, highlighting emerging problems for businesses and academics, as well as new directions for future studies
Roper, S., Du, J., Love, J.H.	Modelling the innovation value chain	Research Policy	2008	Taking into account that innovation events are the culmination of a continuum of information being sourced and transformed, the research analyses a vast circle of companies from Ireland and discovers that complementarity between lateral, forwards, backwards, inner, and public practices of sourcing knowledge is significant. The resulting model may place a premium on the part of a firm's resources such as skills, capital expenditure, and others in the process of value development

(*continued*)

Table 3.3 (continued)

Author(s)	Title	Journal	Year	Key Findings
Hall, B.H., Lotti, F., Mairesse, J.	Innovation and productivity in SMEs: Empirical evidence for Italy	Small Business Economics	2009	Innovation in SMEs exhibits certain unusual characteristics that are not captured by the majority of conventional indices of innovation activity. International competition increases intensity of R&D, particularly for high-technology companies. Among SMEs, it appears as if bigger and more senior companies are not as productive as smaller firms. Innovation of processes and products, and notably innovation of processes, have a favourable effect on the productivity of a firm
Santamaría, L., Nieto, M.J., Barge-Gil, A.	Beyond Formal R&D: Taking Advantage of Other Sources of Innovation in Low- and Medium-Technology Industries	Research Policy	2009	The study discusses how process innovation in low-technology and medium-technology firms may be affected by informal research and development practices and using outer sources. The empirical study is focused on a survey of manufacturing companies representing Spain. The study concludes that architecture, advanced equipment, and training are critical components of understanding any firm's innovation process. These practices have a disproportionately considerable influence on industries of low- and medium-technology

(*continued*)

Table 3.3 (continued)

Author(s)	Title	Journal	Year	Key Findings
Nieto, M.J., Santamaría, L.	Technological collaboration: Bridging the innovation gap between small and large firms	Journal of Small Business Management	2010	The study explores how technical cooperation contributes to the innovation process. The effect of this partnership varies according to the scope of the invention and the type of collaborator. According to the study, vertical cooperation—with suppliers and customers—holds the biggest effect on the innovativeness of a firm. It asserts that such impact is more pronounced for businesses of medium size than for small businesses
Azadegan, A., Dooley, K.J.	Supplier innovativeness, organizational learning styles and manufacturer performance: An empirical assessment	Journal of Operations Management	2010	Suppliers have increased in importance as a source of new technologies and processes. Numerous case studies show how a manufacturer can benefit from supplier innovation. The authors assert that the relationship between supplier innovativeness and output of manufacturer is regulated by the "fit" of the manufacturer's and supplier's learning styles. The findings show that supplier innovation holds a beneficial influence on a variety of dimensions of manufacturer efficiency

(*continued*)

Table 3.3 (continued)

Author(s)	Title	Journal	Year	Key Findings
Chang, C.-H.	The Influence of Corporate Environmental Ethics on Competitive Advantage: The Mediation Role of Green Innovation	Journal of Business Ethics	2011	The study investigates the beneficial impact of ethics that corporations hold regarding the environment on competitive advantage in Taiwan's manufacturing industry using green innovation success as a mediator. Green innovation is classified in the study as innovation of green products and green process innovation. Taiwanese manufacturing firms should improve their environmental stewardship and innovation of green products to boost their advantages comparing to competitors, according to the report

(*continued*)

Table 3.3 (continued)

Author(s)	Title	Journal	Year	Key Findings
Kim, D.-Y., Kumar, V., Kumar, U.	Relationship Between Quality Management Practices and Innovation	Journal of Operations Management	2012	Process management is positively correlated with incremental, radical, and administrative innovation, according to the study. Additionally, the findings indicate that the importance of an individual Quality Management practice is dependent on the value of other Quality Management practices. According to the authors, an organisation's ability to manage processes is crucial for defining routines, creating a learning base, and promoting creative activities. The research analysed data from manufacturing and service firms that were ISO 9001 accredited
Guoyou, Q., Saixing, Z., Chiming, T., Haitao, Y., Hailiang, Z.	Stakeholders' Influences on Corporate Green Innovation Strategy: A Case Study of Manufacturing Firms in China	Corporate Social Responsibility and Environmental Management	2013	Environmental management systems are being integrated into the corporate strategies of an increasing number of companies. Foreign customers are critical in motivating businesses to pursue green process and product innovation strategies. Local and regulatory stakeholders hold no impact on the process of corporate greening

(*continued*)

Table 3.3 (continued)

Author(s)	Title	Journal	Year	Key Findings
Wu, G.-C.	The Influence of Green Supply Chain Integration and Environmental Uncertainty on Green Innovation in Taiwan's IT Industry	Supply Chain Management	2013	The link between green supply chain integration and green innovation is discussed in this paper. Supplier, consumer, and organisational integration all contribute to the advancement of environmentally sustainable products and processes. Demand uncertainty positively moderates each green supply chain integration relation, but technical uncertainty has a negligible moderating impact. The study argues that it is important for management to aim to incorporate skills and resources inside their organisations, suppliers, and customers in order to boost green innovation efficiency. It suggests that managers should actively monitor market dynamics and sustain technical connections between supply chain partners closely

(*continued*)

Table 3.3 (continued)

Author(s)	Title	Journal	Year	Key Findings
Harrison, R., Jaumandreu, J., Mairesse, J., Peters, B.	Does Innovation Stimulate Employment? A Firm-Level Analysis Using Comparable Micro-Data from Four European Countries	International Journal of Industrial Organization	2014	Authors examine the outcome of firm-introduced process and product developments on the growth of employment. The shift in production to new products does not result in a reduction in job requirements. The biggest factor of job growth is the rise in demand for new goods. A maximum of one-third of the net jobs generated by product innovation is projected to be reallocated as a result of business restructuring. However, job growth resulting from business expansion caused by the new product can account for another third of the total

(*continued*)

Table 3.3 (continued)

Author(s)	Title	Journal	Year	Key Findings
Rayna, T., Striukova, L.	From Rapid Prototyping to Home Fabrication: How 3D Printing Is Changing Business Model Innovation	Technological Forecasting and Social Change	2016	A rising view exists that one of the next great technical revolutions will be 3D printing. Despite this, little research has been performed on their effect on business models and creativity in the models of business. In the article, the influence of each step on the essential components of the business model is explored. It demonstrates that 3D printing innovations have the ability to radically alter the way companies innovate by allowing adaptive models of the business
Prajogo, D.I.	The Strategic Fit Between Innovation Strategies and Business Environment in Delivering Business Performance	International Journal of Production Economics	2016	The aim of the article is to look at the position of market environments as variables affecting the effectiveness of innovation strategies. This study demonstrates, using data from 207 Australian manufacturing companies, that diverse settings intensify the product innovation impact on business results. On the flip side, competition indicates a strategic misalignment between product innovation and strategy

(continued)

Table 3.3 (continued)

Author(s)	Title	Journal	Year	Key Findings
Trantopoulos, K., Von Krogh, G., Wallin, M.W., Woerter, M.	External Knowledge and Information Technology: Implications for Process Innovation Performance	MIS Quarterly: Management Information Systems	2017	The incorporation of information from external sources is essential for firm's innovative success. According to the study, firms should align their strategies for acquiring external information with specific IT investments to boost their innovation efficiency. The research was conducted on a nine-year panel of Swiss companies representing a diverse range of manufacturing industries. The results show how, in the modern age, digital transformation will spur innovation
Tang, M., Walsh, G., Lerner, D., Fitza, M.A., Li, Q.	Green Innovation, Managerial Concern and Firm Performance: An Empirical Study	Business Strategy and the Environment	2018	The study shows that when managerial concern for the environment is disregarded, both innovation of green processes and green products substantially positively predict efficiency of firms. However, in the case of inclusion of management concerns, the beneficial impact of green process innovation on firms' efficiency is amplified

(*continued*)

Table 3.3 (continued)

Author(s)	Title	Journal	Year	Key Findings
Najafi-Tavani, S., Najafi-Tavani, Z., Naudé, P., Oghazi, P., Zeynaloo, E.	How Collaborative Innovation Networks Affect New Product Performance: Product Innovation Capability, Process Innovation Capability, and Absorptive Capacity	Industrial Marketing Management	2018	Collaboration will help firms develop their innovation capacities, but only if the focal firm's managers are capable of scanning and gaining external expertise. Collaboration with research organisations and suppliers is critical in the case of process innovation capacity. In accordance with the authors, data obtained in a survey from 258 respondents from Iran's manufacturing industries of high and medium technology shows the value of being cautious when creating innovation networks with a cooperative nature. Cooperation holds a major impact on product innovation only in the presence of absorptive potential

(continued)

Table 3.3 (continued)

Author(s)	Title	Journal	Year	Key Findings
Xie, X., Huo, J., Zou, H.	Green Process Innovation, Green Product Innovation, and Corporate Financial Performance: A Content Analysis Method	Journal of Business Research	2019	In recent years, the business sector has paid increased attention to green technology innovation. However, researchers note that few studies have explored the internal mechanisms that relate innovation of green technology to financial results of a firm. The authors discover that innovation of green processes and green products can help a business boost its profitability. However, the authors assert that the partnership is not moderated by green subsidies

(*continued*)

Table 3.3 (continued)

Author(s)	Title	Journal	Year	Key Findings
Reimann, M., Xiong, Y., Zhou, Y.	Managing a Closed-Loop Supply Chain with Process Innovation for Remanufacturing	European Journal of Operational Research	2019	When products are designed properly, re-manufacturing provides an opportunity to achieve overall sustainability benefits. Only the manufacturer is allowed to innovate in the manufacturing process, while re-manufacturing can be conducted by either the manufacturer or the retailer. A supply chain that is decentralised may be more responsive to process innovation than an integrated supply chain, particularly when the expense of innovating processes is high enough. Inefficiencies caused by making decentralised decisions in a supply chain of a closed-loop can result in both under- and over-investment in process innovation for re-manufacturing

(*continued*)

Table 3.3 (continued)

Author(s)	Title	Journal	Year	Key Findings
Heredia Pérez, J.A., Geldes, C., Kunc, M.H., Flores, A.	New Approach to the Innovation Process in Emerging Economies: The Manufacturing Sector Case in Chile and Peru	Technovation	2019	The study addresses the various approaches to innovation and the various forms of innovation. It includes validation and use of the proposed theoretical model to assess the innovation process by region. In Chile, technical innovation in processes is the major power, while in Peru, non-technological innovation is the dominant power. According to the authors, combining process and organisational innovation improves export efficiency. They identify shortcomings and make recommendations for future studies

(*continued*)

Table 3.3 (continued)

Author(s)	Title	Journal	Year	Key Findings
Zhang, J., Liang, G., Feng, T., Yuan, C., Jiang, W.	Green Innovation to Respond to Environmental Regulation: How External Knowledge Adoption and Green Absorptive Capacity Matter?	Business Strategy and the Environment	2020	The study analysed how regulations impact green product and process innovation in China from the use of data collected in a survey where 237 manufacturing firms took part. The discovered information shows that management and control regulation and regulation based on the market hold a beneficial effect on the acceptance of information from outside. According to the study, green absorptive power reinforces the beneficial effect of market-based control on adoption of knowledge from outside

(*continued*)

Table 3.3 (continued)

Author(s)	Title	Journal	Year	Key Findings
Möldner, A.K., Garza-Reyes, J.A., Kumar, V.	Exploring Lean Manufacturing Practices' Influence on Process Innovation Performance	Journal of Business Research	2020	There is little information how the lean manufacturing practices influence performance of process innovation in manufacturing companies. The study shows that lean methods, both technological and human, hold moderately to strongly positive effect on the feedback and occurrence of process innovation that is gradual and rapid. This, in essence, appears to increase businesses' organisational efficiency as a result of process innovation. The findings dispel the myth that lean manufacturing and innovation cannot coexist peacefully
Dobson, P.W., Chakraborty, R.	Strategic Incentives for Complementary Producers to Innovate for Efficiency and Support Sustainability	International Journal of Production Economics	2020	Not only can manufacturing process innovation improve profitability, but it can also have wider benefits socially and environmentally. People owning businesses may be afraid of investing in innovating processes except if it is driven purely by profit. The authors showed that voluntary agreements backed by government with sector-wide commitments can be beneficial in promoting innovation of processes to promote sustainability and supply chains that are lean

(*continued*)

Table 3.3 (continued)

Author(s)	Title	Journal	Year	Key Findings
Wang, F., Chen, K.	Do Product Imitation and Innovation Require Different Patterns of Organizational Innovation? Evidence from Chinese Firms	Journal of Business Research	2020	Although organisational innovation is viewed as vital for a firm to succeed, its position in the development of a new product is largely unknown. The article aims to examine how product innovation and imitation is affected by organisational innovation in Chinese manufacturing firms. Only a high degree of innovation holds a discernible influence on the intensity of product innovation
Awan, U., Nauman, S., Sroufe, R.	Exploring the Effect of Buyer Engagement on Green Product Innovation: Empirical Evidence from Manufacturers	Business Strategy and the Environment	2021	The study investigates the relationship between buyer-driven information practices, acquisition and combination of resources, and results of product innovation in Pakistani export firms. The findings show that buyer-driven information transfer practices are important for export firms to boost their capacity for absorbing and integrating resources that result in product innovation

Notably, in the 1990s the chief topic of discussion was on the role of efficiency and performance, while the 2000s showed a shift to the importance of product and process innovation in the company's performance. Not surprisingly, in the last few years from 2010, the conversation has gradually shifted to sustainability, green product innovation and how process innovation could be beneficial to the environment.

3.5   A CASE OF INNOVATION: ADDITIVE MANUFACTURING

According to Rayna and Striukova (2016) and Marzi et al. (2018), additive manufacturing is a fairly new technological advancement that attracts increasing attention from manufacturing companies and proves to be a viable avenue for technological innovation in a variety of industries. While additive manufacturing is gaining attention in the manufacturing world, the primary focus of the writings on the subject has been on the technological aspects of the process and from a mostly engineering standpoint (Lee et al., 2017; Dimitrov et al., 2006; Rayna & Striukova, 2016).

Marzi et al. (2018) proposed a case study in Italy's jewel industry that highlighted several benefits but also additional hidden costs of implementing such process innovation to reach a product innovation. The research of Marzi et al. (2018) disclosed that additive manufacturing may be viewed as an innovation of a process that has an extensive impact on a company's profitability, affecting costs, consumer satisfaction and revenues. An increase in prices, an increase in the value provided to the consumer and, eventually, an increase in revenues can be observed.

These effects have an impact on profit generation and, as a result, the competitive advantage of the company. Marzi et al. (2018) show six conceptual themes derived from the latent content study—process innovation, cost savings, consumer value enhancement, sales development, profit growth and sustainable competitive advantage—representing the primary factors affecting the role of additive manufacturing on a company's competitiveness.

The case proposed by the authors demonstrates that additive manufacturing is a major process innovation. Additive manufacturing is often used by companies as the first step in the manufacturing process to produce semi-finished products that serve as the foundation for final production. It becomes apparent how additive manufacturing represents a significant innovation trajectory aimed at improving the enterprise's innovation process both during the prototyping and development phases. With regard to costs, the case analysis demonstrates that the use of additive manufacturing does not result in a significant cost reduction, but rather results in a slight increase because of depreciation, maintenance costs, personnel training costs and, most importantly, raw material costs. Cost structure changes have a major impact on prices. The depreciation of printers, which have a relatively short service life of about two years, has a major effect. As a result, the analysed technology is difficult to handle, as it significantly alters the cost structure of the businesses that implement it. The entrepreneurs who were interviewed stressed the crucial nature of assessing the effect of technology on the company's cost structure. To ensure cost sustainability, this manufacturing process must be continuously enabled, since prolonged inactivity of the system results in a potential loss due to a lack of depreciation. Additive manufacturing machinery also needs considerable maintenance, and staffing costs increase as a result of the requisite preparation for the new technologies and the inclusion of personnel with specialised expertise for additive manufacturing. Finally, the costs of auxiliary material for additive manufacturing are important, as businesses are forced by contractual agreements to purchase them from printer suppliers.

Regarding the effects of additive manufacturing on the value proposition made to customers, it establishes three basic customer service advantages. First, it fosters product innovation, which is consistent with reports on process innovation (Hall et al., 2009; Martinez-Ros, 1999). Indeed, additive manufacturing allows innovative products that are superior in terms of efficiency to be produced. Second, the experience of customers has improved regarding time-to-market and customisation. This has significantly opened up more innovative possibilities for entrepreneurs, and the increased value offered to the consumer has resulted in a desire on the part of consumers to pay higher prices (Marzi et al., 2017).

The effect on revenues is inextricably linked to the previous conceptual trend, as it emerges that additive manufacturing has three primary effects. To begin, revenues increased due to a rise in sales prices as an outcome of the increased value provided to the buyer. Second, the production of new products granted entry into previously untapped market segments. This illustrates how the advent of additive manufacturing enables the penetration of consumer segments that before could not be reached, such as luxury consumers that are very willing to make payments for custom-designed items. Finally, when it comes to the relationship between price and quantity sold, additive manufacturing has primarily affected sales prices, as revenues have risen primarily due to an increase in sales prices occasioned by the improvement in product quality (Baumers & Holweg, 2019).

As a consequence of the three preceding conceptual themes—effects on costs, effects on consumer value, and effects on sales—the case showed that additive manufacturing had a beneficial impact on earnings, which grew as revenues increased while costs remained constant. Additive manufacturing-enabled product innovation results in increased consumer demand, increased willingness to pay and penetration of new segments of the market, leading to an increase in sales and therefore an increase in income, but only if the above-mentioned cost conditions are met (constant use of the 3D printer). The beneficial effect of additive manufacturing on productivity is consistent with the process innovation literature (Reichstein & Salter, 2006). It is worth noting that while the process innovation literature attributes enhanced productivity to cost reduction, the advent of additive manufacturing primarily affects income through sales growth. Finally, in relation to the final conceptual theme, the case studies demonstrated that additive manufacturing can be a source of competitive advantage. However, additive manufacturing proves to be an enduring innovation that provides a sustainable competitive advantage to early adopters (Baumers & Holweg, 2019). In the situation of the jewellery industry, since its implementation by the first competitors, this technology has become a critical component for survival, as consumers now demand goods with a higher degree of customisation, which can only be accomplished by additive manufacturing.

Additive manufacturing as innovation results in a stronger comparative advantage, which is a driving factor in the adoption of additive manufacturing in manufacturing companies. In accordance with Marzi et al. (2018) studies, additive manufacturing drives the competitive advantage, but not sufficiently because it must be combined with other manufacturing technology and entrepreneurship skills to achieve such an advantage.

As a result, it is possible to conclude that additive manufacturing technology represents a viable growth opportunity for the manufacturing industry. Such technology has been shown to be effective in sectors where complex object production is required by lowering production expenses and, in particular, the ability to convert small-scale production to large-scale production (Mellor et al., 2014).

References

Atuahene-Gima, K. (1996a). Differential potency of factors affecting innovation performance in manufacturing and services firms in Australia. *Journal of Product Innovation Management, 13*(1), 35–52.

Atuahene-Gima, K. (1996b). Market orientation and innovation. *Journal of Business Research, 35*(2), 93–103.

Awan, U., Nauman, S., & Sroufe, R. (2021). Exploring the effect of buyer engagement on green product innovation: Empirical evidence from manufacturers. *Business Strategy and the Environment, 30*(1), 463–477.

Azadegan, A., & Dooley, K. J. (2010). Supplier innovativeness, organizational learning styles and manufacturer performance: An empirical assessment. *Journal of Operations Management, 28*(6), 488–505.

Banbury, C. M., & Mitchell, W. (1995). The effect of introducing important incremental innovations on market share and business survival. *Strategic Management Journal, 16*(S1), 161–182.

Baumers, M., & Holweg, M. (2019). On the economics of additive manufacturing: Experimental findings. *Journal of Operations Management, 65*(8), 794–809.

Becheikh, N., Landry, R., & Amara, N. (2006). Lessons from innovation empirical studies in the manufacturing sector: A systematic review of the literature from 1993–2003. *Technovation, 26*(5), 644–664.

Capon, N., Farley, J. U., Lehmann, D. R., & Hulbert, J. M. (1992). Profiles of product innovators among large US manufacturers. *Management Science, 38*(2), 157–169.

Chang, C. H. (2011). The influence of corporate environmental ethics on competitive advantage: The mediation role of green innovation. *Journal of Business Ethics, 104*(3), 361–370.

Cooper, R. G. (2008). Perspective: The stage-gate® idea-to-launch process—Update, what's new, and NexGen systems. *Journal of Product Innovation Management, 25*(3), 213–232.

Damanpour, F. (1991). Organizational innovation: A meta-analysis of effects of determinants and moderators. *The Academy of Management Journal, 34*(3), 555–590.

Damanpour, F. (1996). Organizational complexity and innovation: Developing and testing multiple contingency models. *Management Science, 42*(5), 693–716.

Dimitrov, D., Schreve, K., & de Beer, N. (2006). Advances in three dimensional printing–state of the art and future perspectives. *Rapid Prototyping Journal.*

Dobson, P. W., & Chakraborty, R. (2020). Strategic incentives for complementary producers to innovate for efficiency and support sustainability. *International Journal of Production Economics, 219*, 431–439.

Ettlie, J. E., & Reza, E. M. (1992). Organizational integration and process innovation. *Academy of Management Journal, 35*(4), 795–827.

Evangelista, R., Perani, G., Rapiti, F. and Archibugi, D. (1997). Nature and impact of innovation in manufacturing industry: Some evidence from the Italian innovation survey. *Research Policy, 26*(4), 521–536.

Guoyou, Q., Saixing, Z., Chiming, T., Haitao, Y., & Hailiang, Z. (2013). Stakeholders' influences on corporate green innovation strategy: A case study of manufacturing firms in China. *Corporate Social Responsibility and Environmental Management, 20*(1), 1–14.

Hall, B. H., Lotti, F., & Mairesse, J. (2009). Innovation and productivity in SMEs: Empirical evidence for Italy. *Small Business Economics, 33*(1), 13–33.

Harrison, R., Jaumandreu, J., Mairesse, J., & Peters, B. (2014). Does innovation stimulate employment? A firm-level analysis using comparable micro-data from four European countries. *International Journal of Industrial Organization, 35*, 29–43.

Hatch, N. W., & Mowery, D. C. (1998). Process innovation and learning by doing in semiconductor manufacturing. *Management Science, 44*(11–part-1), 1461–1477.

Herstatt, C., & Von Hippel, E. (1992). From experience: Developing new product concepts via the lead user method: A case study in a "low-tech" field. *Journal of Product Innovation Management, 9*(3), 213–221.

Hurmelinna-Laukkanen, P., Sainio, L. M., & Jauhiainen, T. (2008). Appropriability regime for radical and incremental innovations. *R&D Management, 38*(3), 278–289.

Kim, D. Y., Kumar, V., & Kumar, U. (2012). Relationship between quality management practices and innovation. *Journal of Operations Management, 30*(4), 295–315.

Knudsen, T., & Levinthal, D. A. (2007). Two faces of search: Alternative generation and alternative evaluation. *Organization Science, 18*(1), 39–54.

Koufteros, X., Vonderembse, M., & Jayaram, J. (2005). Internal and external integration for product development: The contingency effects of uncertainty, equivocality, and platform strategy. *Decision Sciences, 36*(1), 97–133.

Laursen, K., & Salter, A. (2006). Open for innovation: The role of openness in explaining innovation performance among UK manufacturing firms. *Strategic Management Journal, 27*(2), 131–150.

Lee, H., Lim, C. H. J., Low, M. J., Tham, N., Murukeshan, V. M., & Kim, Y. J. (2017). Lasers in additive manufacturing: A review. *International Journal of Precision Engineering and Manufacturing-Green Technology, 4*(3), 307–322.

Leydesdorff, L., Carley, S., & Rafols, I. (2013). Global maps of science based on the new web-of-science categories. *Scientometrics, 94*(2), 589–593.

Linder, J. C., Jarvenpaa, S., & Davenport, T. H. (2003). Toward an innovation sourcing strategy. *MIT Sloan Management Review, 44*(4), 43.

Lukas, B. A., & Ferrell, O. C. (2000). The effect of market orientation on product innovation. *Journal of the Academy of Marketing Science, 28*(2), 239–247.

Martinez-Ros, E. (1999). Explaining the decisions to carry out product and process innovations: The Spanish case. *The Journal of High Technology Management Research, 10*(2), 223–242.

Marzi, G., & Caputo, A. (2019). *Responsible entrepreneurship education: Emerging research and opportunities.* IGI Global.

Marzi, G., Dabić, M., Daim, T., & Garces, E. (2017). Product and process innovation in manufacturing firms: A 30-year bibliometric analysis. *Scientometrics, 113*(2), 673–704.

Marzi, G., Zollo, L., Boccardi, A., & Ciappei, C. (2018). Additive manufacturing in SMEs: Empirical evidences from Italy. *International Journal of Innovation and Technology Management, 15*(01), 1850007.

Mathieu, V. (2001). Service strategies within the manufacturing sector: Benefits, costs and partnership. *International Journal of service industry management.*

Mellor, S., Hao, L., & Zhang, D. (2014). Additive manufacturing: A framework for implementation. *International Journal of Production Economics, 149*, 194–201.

Möldner, A. K., Garza-Reyes, J. A., & Kumar, V. (2020). Exploring lean manufacturing practices' influence on process innovation performance. *Journal of Business Research, 106*, 233–249.

Najafi-Tavani, S., Najafi-Tavani, Z., Naudé, P., Oghazi, P., & Zeynaloo, E. (2018). How collaborative innovation networks affect new product performance: Product innovation capability, process innovation capability, and absorptive capacity. *Industrial Marketing Management, 73*, 193–205.

Nieto, M. J., & Santamaría, L. (2007). The importance of diverse collaborative networks for the novelty of product innovation. *Technovation, 27*(6–7), 367–377.

Nieto, M. J., & Santamaría, L. (2010). Technological collaboration: Bridging the innovation gap between small and large firms. *Journal of Small Business Management, 48*(1), 44–69.

Nord, W. R., & Tucker, S. (1987). *Implementing routine and radical innovation.* Lexington Books.

Oldendick, R. W. (2012). Survey research ethics. In *Handbook of survey methodology for the social sciences* (pp. 23–35). Springer.

Pérez, J. A. H., Geldes, C., Kunc, M. H., & Flores, A. (2019). New approach to the innovation process in emerging economies: The manufacturing sector case in Chile and Peru. *Technovation, 79*, 35–55.

Prajogo, D. I. (2016). The strategic fit between innovation strategies and business environment in delivering business performance. *International Journal of Production Economics, 171*, 241–249.

Rayna, T., & Striukova, L. (2016). From rapid prototyping to home fabrication: How 3D printing is changing business model innovation. *Technological Forecasting and Social Change, 102*, 214–224.

Reichstein, T., & Salter, A. (2006). Investigating the sources of process innovation among UK manufacturing firms. *Industrial and Corporate Change, 15*(4), 653–682.

Reimann, M., Xiong, Y., & Zhou, Y. (2019). Managing a closed-loop supply chain with process innovation for remanufacturing. *European Journal of Operational Research, 276*(2), 510–518.

Rogers, M. (2004). Networks, firm size and innovation. *Small Business Economics, 22*(2), 141–153.

Roper, S., Du, J., & Love, J. H. (2008). Modelling the innovation value chain. *Research Policy, 37*(6–7), 961–977.

Santamaría, L., Nieto, M. J., & Barge-Gil, A. (2009). Beyond formal R&D: Taking advantage of other sources of innovation in low- and medium-technology industries. *Research Policy, 38*(3), 507–517.

Sirilli, G., & Evangelista, R. (1998). Technological innovation in services and manufacturing: Results from Italian surveys. *Research Policy, 27*(9), 881–899.

Smith, W. K., & Tushman, M. L. (2005). Managing strategic contradictions: A top management model for managing innovation streams. *Organization Science, 16*(5), 522–536.

Tang, M., Walsh, G., Lerner, D., Fitza, M. A., & Li, Q. (2018). Green innovation, managerial concern and firm performance: An empirical study. *Business Strategy and the Environment, 27*(1), 39–51.

Terziovski, M. (2010). Innovation practice and its performance implications in small and medium enterprises (SMEs) in the manufacturing sector: A resource-based view. *Strategic Management Journal, 31*(8), 892–902.

Thomke, S., & Von Hippel, E. (2002). Customers as innovators: A new way to create value. *Harvard Business Review, 80*(4), 5–12.

Trantopoulos, K., von Krogh, G., Wallin, M. W., & Woerter, M. (2017). External knowledge and information technology: Implications for process innovation performance. *MIS Quarterly, 41*(1), 287–300.

Van de Vrande, V., De Jong, J. P., Vanhaverbeke, W., & De Rochemont, M. (2009). Open innovation in SMEs: Trends, motives and management challenges. *Technovation, 29*(6–7), 423–437.

Van Eck, N. J., & Waltman, L. (2010). Software survey: VOSviewer, a computer program for bibliometric mapping. *Scientometrics, 84*(2), 523–538.

Van Eck, N. J., & Waltman, L. (2014). Visualizing bibliometric networks. In *Measuring scholarly impact* (pp. 285–320). Springer International Publishing.

Van Eck, N. J., Waltman, L., Dekker, R., & van den Berg, J. (2010). A comparison of two techniques for bibliometric mapping: Multidimensional scaling and VOS. *Journal of the American Society for Information Science and Technology, 61*(12), 2405–2416.

Waltman, L., van Eck, N. J., & Noyons, E. C. (2010). A unified approach to mapping and clustering of bibliometric networks. *Journal of Informetrics, 4*(4), 629–635.

Wang, F., & Chen, K. (2020). Do product imitation and innovation require different patterns of organizational innovation? Evidence from Chinese firms. *Journal of Business Research, 106*, 60–74.

Wu, G. C. (2013). The influence of green supply chain integration and environmental uncertainty on green innovation in Taiwan's IT industry. *Supply Chain Management: An International Journal.*

Xie, X., Huo, J., & Zou, H. (2019). Green process innovation, green product innovation, and corporate financial performance: A content analysis method. *Journal of Business Research, 101*, 697–706.

Zhang, J., Liang, G., Feng, T., Yuan, C., & Jiang, W. (2020). Green innovation to respond to environmental regulation: How external knowledge adoption and green absorptive capacity matter? *Business Strategy and the Environment, 29*(1), 39–53.

CHAPTER 4

A Review of Excessive Development Studies on the New Product Development Process

Abstract The inclination to design novel products and services excessively is a serious threat to the new product development process. Excessive development manifests itself in a variety of ways, with different outcomes. Nonetheless, while the problem of excessive development is well known in the new product development practices, there is a fragmented scientific debate around this concept. This chapter explores the evolution of the studies about excessive development over the time, tracing the debate around such a multifaceted phenomenon.

Keywords New product development • Uncertainty • Fuzzy • Excessive development

4.1 THE PROBLEM OF NEW PRODUCT DEVELOPMENT AND UNCERTAINTY

Companies may gain a competitive advantage in the market by developing innovative technologies, providing new services and expanding their product range. However, the growing need for product differentiation, constant technical advancements and rapid shifts in consumer tastes often compelled businesses to create and propose products that exceeded the necessities of the industry and end users (Birkinshaw, 2018; Graham & Senge, 1980; Karlsson & Ahlström, 1996). Many anecdotal cases of

G. Marzi, *Uncertainty-driven Innovation*,
https://doi.org/10.1007/978-3-030-99534-8_4

excessive development can be found in daily life. For example, the BMW series 7 featured the iDrive device, which offered approximately 700 features through displays with multifunctional properties and multistep operations. Because the iDrive method is so sophisticated, the motor vehicle manufacturer was compelled to provide a user's manual in case the car was taken by a valet parker (Niedermaier et al., 2009). Rust et al. (2006) also discussed the difficulty of using the Bosch Benvenuto B30 coffee machine, which features a plethora of choices for users who "only want a coffee fast".

Thus, different types of excessive development are characterised as an adverse condition of the product and the new product development project that has been labelled as "Over Featuring" by the author of this book in Marzi (2022). Over Featuring refers to any form of excessive development that exceeds the requirements of customers, the market, the plans and the possibilities that stem from the resources of the company (Marzi, 2022), in connection with the previous studies of Coman and Ronen (2010) and Shmueli and Ronen (2017). In the present book "Over Featuring" and "excessive development" are used as synonyms.

The importance of the topic is clear if even the American National Aeronautics and Space Administration (NASA) cautioned in 1992 about incorporating unnecessary functionality during the project specification process, listing it as one of the top new product development threats (NASA Goddard Space Flight Center, 1992). However, Over Featuring have been analysed mostly in the context of software development, likely because software development's rapid development cycles facilitate the proliferation of Over Featuring and provide a suitable sample for the analysis. Nonetheless, the idea of Over Featuring has received scant academic and practitioner attention, especially in the realms of physical objects (Marzi, 2022).

The existing literature sheds light on a few fragments of such a tangled forest, but only provides a partial description of the phenomena.

4.2   An Overview of Excessive Development on the New Product Development Process

In 1996, Christensen and Bower (1996) investigated the global disc drive industry, finding that businesses, particularly the leading ones, proposed highly performing drivers with more capabilities than the final users needed. Excessive development, according to the writers, is caused by strong competitiveness in the technology industry, a need to place goods on high-tier markets and the need to predict technical developments in the mainstream market.

Schmidt et al. (2001) proposed a Delphi analysis five years later to classify the most serious threats to software projects. The types of excessive development are included in 2 of the 14 groups of defined risks. The first group reflects on the project's scale, with reasons for excessive growth including undefined scope, repeated scope shifts, scope creep, unclear users' desires and excessive organisational units taking part in ventures. The second group deals with specifications, demonstrating how a project's loss can be caused by a shortage of requirement "freezing" during the development process with unfamiliar requirements (Bayus, 2013; Birkinshaw, 2018). During the years, scope creep received vast attention from scholars and practitioners, as it is one of the clearly emergent and easily identified types of Over Featuring. Exemplary investigations on the scope creep topic can be found in the studies of Chen et al. (2009), Knight and Robinson Fayek (2002) and Nilsson-Witell et al. (2005).

Thompson et al.'s contribution in 2005 gave Over Featuring a more formal nature. The authors looked at how extra features might improve the incentive to buy a product, although many features can daunt customers. Thompson et al. (2005) coined the term "feature fatigue" to describe the difficulty and subsequent disappointment of users by utilising an overly complicated app. The authors also discussed how a variety of features might influence the initial purchase, repurchase or net present value of a security. Despite Thompson et al.'s (2005) extremely important contribution, the study only identified user exhaustion during the usage of overly complicated items without allowing the new product development mechanism. In addition, Damian and Chisan (2006) investigated how requirement management in a software development project may be critical for competitiveness, efficiency and project risks. "Feature creep" is recognised among many as one of the leading causes of new product development project failure.

In 2007, Elliott's intervention (Elliott, 2007) attempted to investigate some of the causes of feature creep and how to prevent it, paving the groundwork for further literature developments beyond the computing sphere. Feature creep, according to Elliott (2007), stems from the inherent need to change and invent goods by responding to consumer, sales and marketing department demands.

Ronen and Pass (2008) made another important move in exploring the forms of excessive development in 2008, when they described, from the new product creation viewpoint, the propensity to create new products and services with unnecessary features as overspecification and overdesign. Following that, in 2010, Coman and Ronen (2010) expanded on the idea of overspecification and overdesign by presenting several case studies that show the negative consequences of such pathologies. The authors have looked at the mechanisms behind such pathologies, finding that over-specification and overdesign are more often caused by interactions between departments (usually R&D and marketing) with hazy and undefined perceptions about what produces value in a product or service at various stages of the development phase. The cases described by Coman and Ronen (2010) are helpful in understanding the outcomes of Over Featuring. The authors cited, among others, the example of Microsoft, which bought a forward-thinking start-up focused on computer security. The data protection software intended to penetrate the industry was drastically postponed. Microsoft Word includes a secret pinball application, and Microsoft Excel includes a stealthy flight simulator (Coman & Ronen, 2010). Johann Fust, a lawyer, lent money to Gutenberg, and Gutenberg incorporated Johann Fust into his firm. The perfectionism that Gutenberg possessed—the need to increase the standard and add features such as the ability of colour printing—ultimately led to his bankruptcy (Coman & Ronen, 2010). The redesign phase for the Bradley tank lasted over two decades and as a result a slew of new capabilities were added to the Bradley, including a rocket carrier, amphibious capabilities and gun systems (Coman & Ronen, 2010).

Simultaneously, Gill (2008) investigated what features bring value to consumers and the benefits of offering convergent goods. The author compared the utilitarian versus hedonic consumer base and utilitarian versus hedonic functionalities, concluding that the addition of a hedonic feature (such as a camera) to utilitarian goods (such as cell phones) is welcomed. Users, on the other hand, are more resistant to the addition of utilitarian functions the more hedonistic a product is (Gill, 2008). Han et al. (2009) investigated customers' preferences for converged or dedicated goods. Consumers choose convergent goods where the technical sophistication of the product is minimal, according to the findings. Consumers favour dedicated goods where technical sophistication and efficiency are strong.

Following that, Buschmann (2009, 2010) suggested a list of the most popular and harmful software development problems, including scope creep, "Featuritis", "Flexibilitis" and "Performitis". Buschmann (2010) connects project managers' erratic actions with ineffective project management activities as one of the triggers of excessive development. In the case of Enterprise resource planning (ERP) execution, Chen et al. (2009) demonstrated how an infective project management mechanism contributes to failure. Scope creep has been described as one of the most significant reasons for project failures. Customers, engineers and project managers can be seduced by a system that accommodates one-size-fits-all, which will cover all the company's processes, exposing the whole project to a large probability of failure, poor usability and unnecessary difficulty. In a similar way, Choi and Bae (2009) demonstrate how shifting users' requirements will result in unanticipated and unregulated requirements creep for long-term and large-scale ventures.

Thompson and Norton (2011) looked at the societal implications of choosing a product with a large range of features and capabilities in 2011. The findings revealed that products with a large range of attributes and functionalities encourage others to provide a good social view of them. Users can tolerate a reduction in efficiency as a result of an unnecessary number of functions in exchange for a positive social impression. These results are consistent with Gill's previous research on hedonic versus functional goods (Gill, 2008). In reality, according to Thompson and Norton (2011), customers favour feature-rich items for public display when it comes to conspicuous consumption (e.g. luxury goods). Consumers, on the other hand, want the least featured brands when searching for performance and accessibility (Schreier et al., 2012).

Bjarnason et al. (2012) investigated the drivers, sources and consequences of another form of excessive development affecting software development programmes, overscoping, in their study of software release management. The results comply with other Over Featuring forms: consistency issues, inability to satisfy consumer demands, production delays and final product usability issues.

Since 2013, a community of researchers has been empirically studying the behavioural origins of excessive development. Belvedere et al. (2013) suggested a study of the cognitive biases that lead to overdesign. The authors discovered several cognitive biases involved in the generation of overdesign using a group of industrial engineers. Overdesign is fuelled by an unrealistic selection of functionalities, developers' excessive overconfidence in the project's optimistic result and a misleading assessment of the economic return of additional product functionality, according to the study.

In two other laboratory trials, Shmueli et al. (2015, 2016) built on the function of behavioural variables in producing excessive development. In an experiment involving about 200 students, Shmueli et al. (2015) investigated the function of emotional participation in generating overspecification in a software development context. Participants who spent cognitive activity defining features that were not required for the overall framework overvalued certain features compared to the importance given to them prior to the specification phase, according to the findings. The second research by Shmueli et al. (2016) included another trial of 100 students to evaluate the effect of an outside-view method (looking at related scenarios instead of focusing on the particular case before formulating a potential decision) in eliminating unreasonable behaviours in software development ventures, such as time underestimation, scope saturation and overspecification. To prevent excessive development and cognitive biases, the study was required to predict development times and the number of features using two exterior-view approaches: reference knowledge regarding previous times of delivery and perspective of the role. The findings revealed that software development programmes suffer from overspecification, overscoping and time underestimation, with the outside-view method that is able to minimise but not eradicate excessive development practices and their related biases (Choi et al., 2011).

Li et al. (2014) and Wu et al. (2015) suggested two studies to reduce feature fatigue. The first research (Li et al., 2014) proposed a continuous fuzzy Kano model-based method for reducing unneeded functionality by considering the imprecision and complexity of consumer requirements. To assist developers in recognising unneeded functionality, a feature fatigue index was proposed to assess each feature's contribution to feature fatigue. A multimethod model was suggested in the second analysis (Wu et al., 2015) to maximise the number of function combinations. To find Pareto-optimal solutions for choosing the optimal number of features in a product, the grey model, worst-case scenario and non-dominated sorting genetic algorithm II were used.

Finally, in 2017, Shmueli and Ronen (2017) proposed a literature review to better understand excessive development in the context of software development. The authors discovered many overlapping concepts used in software development analysis and practice. Beyond needs, beyond plans and beyond resources are three macro-compartments developed by scholars to classify those concepts. The fragmented literature on excessive development in software production is best understood and systematised using these definitions.

A significant number of observational findings published in 2019 contributed to a greater understanding of Over Featuring. Bianchi et al. (2019) proposed a study of 307 app developers to further explore the function of cognitive biases and personality characteristics in Over Featuring. Bianchi et al. (2019) studied the effect of developers' perceptual, rational and behavioural characteristics on excessive development using the tripartite model of human attitudes. The results revealed that emotional attachment is linked to excessive development. A logical cognitive style is more concerned with designing one-size-fits-all apps, while a style that is intuitive cognitive drives the addition of supplementary functionality to the current scope of the product.

De Giovanni (2019) investigated how manufacturers' tactics might help reduce feature fatigue caused by overspecified devices. To mitigate feature fatigue, manufacturers may recommend either a joint programme with retailers (paying a fraction of the retailer's store facilitator, such as help ads and/or promotional strategies) or an ad hoc facilitator strategy (e.g. special training packages, web channels, tablets, home utilities, free upgrades and support services). The findings revealed that all supply chain participants are interested in working together to boost profits and diminish the feature fatigue effects that can be detrimental. Cooperative systems help to decrease harmful feature fatigue effects, but they do not necessarily increase earnings. Ad hoc facilitators have the reverse benefit; they do not necessarily minimise the detrimental feature fatigue effect, but they help supply chain participants to optimise earnings.

Garcia et al. (2019) used a naval vessel design case study to show the dangers of overspecification and the importance of stakeholders' preferences in shaping the incorporation of unnecessary features in a project. Scholars have shown the negative impact of overlapping specification demands from various vessel designs and requirements. Garcia et al. (2019) suggested a range of tools, including alignment A, B, degree of misalignment (DoM) and individual expectation fulfilment index (IEFI), to assist stakeholders in understanding overspecification problems and reducing the likelihood of project failure.

Then, utilising several cases of medium-sized knitwear firms, Gregori and Marcone (2019) demonstrated that overdesign and overspecification are inherent phenomena in any innovation project, new product development project and R&D operation. Despite the detrimental effects of overdesign and overspecification on business systems, Over Featuring cannot be entirely eliminated from the operations of developing a new product because the mechanism of developing a new product itself is fraught with ambiguity. To mitigate the harmful effects of Over Featuring, scholars recommend two countermeasures: variation reduction systems and early specification freezing.

Jain (2019) used hyperbolic discounting to study user biases and feature fatigue. Hyperbolic discounting means that customers buy goods with features they do not use, overestimating the importance of certain features at the point of purchasing, while applying lower utility to certain features while they are used. As a result, marketers aim to sell goods with numerous features to maximise the initial sales cost, causing consumers to experience feature fatigue during use. Jain (2019) suggested a game-theoretic model that simulates the need for users to spend more time learning about additional product features. Companies should concentrate their efforts on making learning easier instead of concentrating on investments in new functions, according to hyperbolic discounting.

Komal et al. (2020) investigated the causes of scope creep in software development ventures by empirically demonstrating that they are linked to three key dimensions, namely technology, organisation and human. De Giovanni (2020) studied the effect of competitors and supply chain partners on firms' ability to produce feature-based goods by analysing consumers' customisation choices and the related organisational plans to circumvent the product confusion caused by features. The author illustrated that feature-based development capabilities are highly affected by the decisions of rivals and supply chain partners when servitisation is a required precondition for maintaining an option of valuable customisation. The study shows how advanced forecasting programmes, vendor-controlled inventory and fulfilment demand preparation will help businesses maximise their efficiency. Finally, Shabi et al. (2021) proposed a decision support model for defining and handling overspecification in large-scale device construction programmes. The author suggests an optimisation dilemma of project restrictions to minimise the negative effects of overspecification, as well as a paradigm for handling overspecification in major projects based on an estimate of the expense and benefit to the consumer associated with each design requirement.

REFERENCES

Bayus, B. L. (2013). Crowdsourcing new product ideas over time: An analysis of the Dell IdeaStorm community. *Management Science, 59*(1), 226–244.

Belvedere, V., Grando, A., & Ronen, B. (2013). Cognitive biases, heuristics, and overdesign: An investigation on the unconscious mistakes of industrial designers and on their effects on product offering. In *Behavioral issues in operations management: New trends in design, management, and methodologies* (Vol. 9781447148, pp. 125–139). Springer.

Bianchi, M., Marzi, G., Zollo, L., & Patrucco, A. (2019). Developing software beyond customer needs and plans: An exploratory study of its forms and individual-level drivers. *International Journal of Production Research, 57*(22), 7189–7208.

Birkinshaw, J. (2018). What to expect from agile. *MIT Sloan Management Review, 59*(2), 39–42.

Bjarnason, E., Wnuk, K., & Regnell, B. (2012). Are you biting off more than you can chew? A case study on causes and effects of overscoping in large-scale software engineering. *Information and Software Technology, 54*(10), 1107–1124.

Buschmann, F. (2009). Learning from failure, part 1: Scoping and requirements woes. *IEEE Software, 26*(6), 68–69.

Buschmann, F. (2010). Learning from failure, part 2: Featuritis, performitis, and other diseases. *IEEE Software, 27*(1), 10–11.

Chen, C. C., Law, C. C. H., & Yang, S. C. (2009). Managing ERP implementation failure: A project management perspective. *IEEE Transactions on Engineering Management, 56*(1), 157–170.

Choi, J. N., Sung, S. Y., Lee, K., & Cho, D. S. (2011). Balancing cognition and emotion: Innovation implementation as a function of cognitive appraisal and emotional reactions toward innovation. *Journal of Organizational Behavior, 32*(1), 107–124.

Choi, K. S., & Bae, D. H. (2009). Dynamic project performance estimation by combining static estimation models with system dynamics. *Information and Software Technology, 51*(1), 162–172.

Christensen, C. M., & Bower, J. L. (1996). Customer power, strategic investment, and the failure of leading firms. *Strategic Management Journal, 17*(3), 197–218.

Coman, A., & Ronen, B. (2010). Icarus' predicament: Managing the pathologies of overspecification and overdesign. *International Journal of Project Management, 28*(3), 237–244.

Damian, D., & Chisan, J. (2006). An empirical study of the complex relationships between requirements engineering processes and other processes that lead to payoffs in productivity, quality, and risk management. *IEEE Transactions on Software Engineering, 32*(7), 433–453.

De Giovanni, P. (2019). A feature fatigue supply chain game with cooperative programs and ad-hoc facilitators. *International Journal of Production Research*, *57*(13), 4166–4186.

De Giovanni, P. (2020). When feature-based production capabilities challenge operations: Drivers, moderators, and performance. *International Journal of Operations and Production Management*, *40*(2), 221–242.

Elliott, B. (2007). Anything is possible: Managing feature creep in an innovation rich environment. *IEEE International Engineering Management Conference*, *2007*, 304–307.

Garcia, J. J., Pettersen, S. S., Rehn, C. F., Erikstad, S. O., Brett, P. O., & Asbjørnslett, B. E. (2019). Overspecified vessel design solutions in multi-stakeholder design problems. *Research in Engineering Design*, *30*(4), 473–487.

Gill, T. (2008). Convergent products: What functionalities add more value to the base? *Journal of Marketing*, *72*(2), 46–62.

Graham, A. K., & Senge, P. M. (1980). A long-wave hypothesis of innovation. *Technological Forecasting and Social Change*, *17*(4), 283–311.

Gregori, G. L., & Marcone, M. R. (2019). R&D and manufacturing activities regarding managerial effectiveness and open strategy: An industry focus on luxury knitwear firms. *International Journal of Production Research*, *57*(18), 5787–5800.

Han, J. K., Chung, S. W., & Sohn, Y. S. (2009). Technology convergence: When do consumers prefer converged products to dedicated products?. *Journal of Marketing*, *73*(4), 97–108.

Jain, S. (2019). Time inconsistency and product design: A strategic analysis of feature creep. *Marketing Science*, *38*(5), 835–851.

Karlsson, C., & Ahlström, P. (1996). The difficult path to lean product development. *Journal of Product Innovation Management*, *13*(4), 283–295.

Knight, K., & Robinson Fayek, A. (2002). Use of fuzzy logic for predicting design cost overruns on building projects. *Journal of Construction Engineering and Management*, *128*(6), 503–512.

Komal, B., Janjua, U. I., Anwar, F., Madni, T. M., Cheema, M. F., Malik, M. N., & Shahid, A. R. (2020). The impact of scope creep on project success: An empirical investigation. *IEEE Access*, *8*, 125755–125775.

Li, M., Wang, L., & Wu, M. (2014). An integrated methodology for robustness analysis in feature fatigue problem. *International Journal of Production Research*, *52*(20), 5985–5996.

Marzi, G. (2022). On the nature, origins and outcomes of Over Featuring in the new product development process. *Journal of Engineering and Technology Management*, *64*, 101–685.

NASA Goddard Space Flight Center. (1992). Recommended approach to software development revision 3. In *Software engineering laboratory series* (Issue June). https://ntrs.nasa.gov/archive/nasa/casi.ntrs.nasa.gov/19930009672.pdf

Niedermaier, B., Durach, S., Eckstein, L., & Keinath, A. (2009). The new BMW iDrive–applied processes and methods to assure high usability. In *International conference on digital human Modeling* (pp. 443–452). Springer.

Nilsson-Witell, L., Antoni, M., & Dahlgaard, J. J. (2005). Continuous improvement in product development: Improvement programs and quality principles. *International Journal of Quality & Reliability Management, 22*(8), 753–768.

Ronen, B., & Pass, S. (2008). *Focused operations management: Achieving more with existing resources.* John Wiley & Sons.

Rust, R. T., Thompson, D. V., & Hamilton, R. W. (2006). Defeating feature fatigue. *Harvard Business Review, 84*(2), 37–47.

Schmidt, R., Lyytinen, K., Keil, M., & Cule, P. (2001). Identifying software project risks: An international Delphi study. *Journal of Management Information Systems, 17*(4), 5–36.

Schreier, M., Fuchs, C., & Dahl, D. W. (2012). The innovation effect of user design: Exploring consumers' innovation perceptions of firms selling products designed by users. *Journal of Marketing, 76*(5), 18–32.

Shabi, J., Reich, Y., Robinzon, R., & Mirer, T. (2021). A decision support model to manage overspecification in system development projects. *Journal of Engineering Design*, 1–23.

Shmueli, O., Pliskin, N., & Fink, L. (2015). Explaining over-requirement in software development projects: An experimental investigation of behavioral effects. *International Journal of Project Management, 33*(2), 380–394.

Shmueli, O., Pliskin, N., & Fink, L. (2016). Can the outside-view approach improve planning decisions in software development projects? *Information Systems Journal, 26*(4), 395–418.

Shmueli, O., & Ronen, B. (2017). Excessive software development: Practices and penalties. *International Journal of Project Management, 35*(1), 13–27.

Thompson, D. V., Hamilton, R. W., & Rust, R. T. (2005). Feature fatigue: When product capabilities become too much of a good thing. *Journal of Marketing Research, 42*(4), 431–442.

Thompson, D. V., & Norton, M. I. (2011). The social utility of feature creep. *Journal of Marketing Research, 48*(3), 555–565.

Wu, M., Wang, L., Long, H., & Li, M. (2015). Feature fatigue analysis in product development. *Total Quality Management and Business Excellence, 26*(1–2), 218–232.

CHAPTER 5

Managing Uncertainty Through Stage-Gate, Agile and Overspecification

Abstract The majority of innovation projects fall short of their quality, speed and cost objectives. One important explanation for this observation is the relevant degree of complexity found in settings of innovation. The traditional linear Stage-Gate models aim to mitigate ambiguity by rigorous preparation to prevent feature changes at the last minute. The essence of more modern Agile models is to adjust products iteratively to volatility, even during the final stages of development. In the middle, Overspecification aims to prevent uncertainty by adding additional features to prevent future adjustments. This research examines the impact of Agile and Stage-Gate models, together with Overspecification strategies, on the new product development process from an uncertainty management viewpoint.

Keywords New product development • Stage-Gate • Agile • Overspecification

5.1 AGILE AND STAGE-GATE MODELS ON THE NEW PRODUCT DEVELOPMENT PROCESS

In recent years, firms' R&D investment has been increasing progressively, showing that innovation is becoming more relevant than ever, topping managers' corporate agenda (Choi & Valikangas, 2001; Evanschitzky et al., 2012). Despite this, nearly half of new product development

© The Author(s), under exclusive license to Springer Nature   73
Switzerland AG 2022
G. Marzi, *Uncertainty-driven Innovation*,
https://doi.org/10.1007/978-3-030-99534-8_5

programmes fall short of their time, cost and efficiency objectives. Contrary to expectations, and despite unparalleled technical advancements, the market output of new technologies has seen stability for five decades, at average levels, when compared with other industry processes (Lee & Markham, 2016; Michaelis et al., 2018).

The academic analysis in the literature on product innovation describes many variables that contribute to superior new product development efficiency, including technical, process and operational architecture considerations, and product and market attributes (Cooper, 2014; Ernst, 2002; Michaelis et al., 2018; Montoya-Weiss & Calantone, 1994).

Recently, a new class of iterative and flexible process models has arisen that function differently from traditional linear systems. The linear systems, which involve Stage-Gate, require the front-end development of comprehensive product requirements and schedules, the sequential organisation of project stages and implementing plans in accordance with negotiated specifications, a stringent workflow and precisely specified guidelines for passing successive gates (Cooper et al., 2002; Ettlie & Elsenbach, 2007; Ettlie & Rosenthal, 2011; Karlstrom & Runeson, 2005). On the contrary, models that have flexibility, for example, the Agile method, encourage very little upfront planning, early validation of customers through swift prototyping and regular testing, and the organisation of development work in repetitions of several test cycles (Beck et al., 2001; Birkinshaw, 2018; Ries, 2011).

A growing body of research shows how companies are rapidly embracing Agile and scalable frameworks, citing the advantages of improved team efficiency, waste savings, compressed project lead times and higher and more reliable consistency (Birkinshaw, 2018; Thomke, 1997). However, scholars have stressed the complexities and difficulties associated with applying linear Stage-Gate methodologies, specifically in very unpredictable and complex project environments (Sommer et al., 2015; Karlstrom & Runeson, 2005; Ward & Sobek II, 2014). Despite growing scholarly interest, scientific research on the implications of Stage-Gate versus Agile remains sparse, relying primarily on illustrative case studies (Sommer et al., 2015).

The crucial challenge of innovation is to handle the intrinsic complexity inherent in creating something new effectively (Sull, 2004). Uncertainty arises from both exogenous and endogenous factors, such as changing consumer tastes, competitor strategic steps and nascent technological trajectories, as well as endogenous sources, such as developers' innovative approaches to project-related findings or the co-evolution of technical solutions in the interaction of product components, as has been shown the various forms of Over Featuring (Marzi, 2022).

The linear Stage-Gate and iterative Agile models represent two radically dissimilar techniques for managing complexity. Through extensive research and preparation at the beginning of the innovation process, linear Stage-Gate attempts to manage complexity. The foundation is to spend significant capital on acquiring early knowledge, for example, industry intelligence and application scouting, with the intention of creating a reliable design for the product and business case, thus keeping the probability of expensive updates and adjustments in the next stages of project to a minimum (Antons et al., 2019; Birkinshaw, 2018; Iansiti, 1995). Agile models, on the other hand, aim to react to instability by evolutionarily adapting product requirements and plans. According to these models, significant advance investment in estimating and describing concepts gives a small return because technological and business circumstances will alter radically and unpredictably during the time of the project.

While distinctions exist between the two frameworks, recent literature shows that companies are combining Agile strategies with Stage-Gate framework components. The first results confirm the compatibility of both models and show that companies that use a blended solution have a higher success rate within new product development projects (Antons et al., 2019; Cooper & Sommer, 2016; Sommer et al., 2015).

This research examines the implications of a specific solution to handling the inherent uncertainty in innovation projects by overspecification (Coman & Ronen, 2010). As mentioned in the previous chapter, overspecification entails having "security margins" in the current product's specification and/or design by "Defining product or service specifications beyond the actual needs of the customer or the market" (Ronen & Pass, 2008, p. 162). This redundancy, which may take several forms, such as additional functionality or unnecessary technical efficiency, may help prevent loopbacks if the project's climate changes over the course of development. Overspecification is an attempt to protect against volatility by including margins in product specifications and keeping more possibilities available. Unfortunately, overspecification is far from inexpensive in terms of financial resources, development time and risk associated with possible feature fatigue from the final users (Thompson et al., 2005; Buschmann, 2010; Coman & Ronen, 2010; Bjarnason et al., 2012) and could configure more as a weakness than a helpful strategy for new product development projects.

5.2   How Agile and Stage-Gate Models Manage the Uncertainty of the New Product Development Process

Innovation practice can be viewed as a scheme for managing uncertainty (Bayus, 2013). Development activities, for example, analysing, engineering and testing, permit the transformation of information that is not complete into knowledge that is useful for the designing of new products (Bayus, 2013; Loch et al., 2011).

Not all innovation potentials created in the fuzzy front-end are pursued further, but filters occur at various points in the process that exclude ineffective ideas and variations of design criteria, allowing capital to be concentrated on the most promising ones (Schreier et al., 2012; Terwiesch & Ulrich, 2009). Developers choose these testing choices with only a limited future understanding. By having an interpretation of the innovation process like this, it is possible to compare the models of Stage-Gate and Agile in terms of the structure of the knowledge production process, the pacing and how convergence is approached, and, in general, how the intrinsic complexity of creative endeavours is dealt with. These distinctions can also help in understanding the way such models, including overspecification strategies, affect the efficiency, cost and quality or how well the innovation projects operate in complex and unpredictable environments.

Since the 1970s, Stage-Gate has become the de facto template for handling the modern product development process (Cooper et al., 2002). They recommend segmenting development work into sequential phases, such as concept specification, architecture, execution, testing and launching, separated by gates and decision points. A critical condition in Stage-Gate models is that only after the preceding phase is finished and passed through the gate's official inspection should working on a downstream phase begin.

To attain the optimum mix, developers conduct rigorous knowledge collection practices, such as market analysis and technical intelligence, in order to establish a balance between consumer demands and technical capabilities and to build an appealing business case. The design freeze step of a project is the stage at which senior project management accepts the final package of product requirements and transfers them to a plan with no changes approved during the rest of the development period. However, this critical moment emerges early in Stage-Gate models, where the degree of demand and technological volatility is strong. Cognitive biases, for instance, fallacy in planning and thinking wishfully limit the precision of forecasts set by developers under these circumstances (Bayus, 2013; Kahneman & Tversky, 1977). Additionally, the prolonged time period between an expected design freeze and product release raises the probability that new relevant material, produced in internal and external ways, arises and demonstrates that the chosen concept no longer provides the optimum product-market match.

The inflexible strategy entails missing important design development tools and could result in introducing a lower quality product with an outdated technological foundation into a market where consumer needs have long changed (Schreier et al., 2012; Krishnan et al., 1997; Krishnan & Zhu, 2006). Because of the linear and sequential nature of Stage-Gate's process architecture, these changes take the form of costly and time-consuming loopbacks, through which engineers are required to repeat previous stages and get the approach reworked before it is converged into a suitable design. It is not guaranteed that this will happen, since an alteration of a single variable will cause modifications to other components, potentially resulting in uncertainty (Bayus, 2013; Schreier et al., 2012; Ward & Sobek II, 2014).

As a result of this ambiguity, Stage-Gate models are incapable of dealing with the unavoidable changes that exist in complex and unpredictable new product development settings. Hence:

Hypothesis 5.1
In unstable and complex settings, Stage-Gate model–based management of the new product development process is linked to lower quality, speed and cost performance.

Agile is a generic term that stands for a group of iterative development methodologies. Agile started in 2001, as, at this time, some of the key practising specialists drafted a manifesto of principles and guidance aimed at streamlining the process of creating a new code for software. However, previous approaches, for instance, the spiral paradigm for development, proposed several of the concepts that underpin Agile (Birkinshaw, 2018; Boehm, 1991; Karlsson & Ahlström, 1996; Liker & Morgan, 2006).

The nature of Agile is responding to transition, and developers who use these models anticipate significant changes in consumer expectations, technical capabilities and other environmental factors over the course of the project's existence (Birkinshaw, 2018). Thus, the aim is to leverage this useful knowledge and have it integrated into the emerging design of the product to attain a fit that is better for the market without sacrificing the efficiency of the costs or time it takes for the product to evolve from idea to becoming available on the market. This involves a method that contrasts linear plan–based models, as well as a management framework that enables the milestone of the concept freeze to be deferred up to the very latest probable instant prior to consumer launch (Cao et al., 2009).

Agile methodology is grounded in a variety of concepts. One is the incremental and iterative product development cycle (Bjarnason et al., 2012; Cao et al., 2009). These cycles are condensed into a series of time-limited task sessions or sprints. Sprints usually last one or a few weeks and cannot extend the specified time. This lays a healthy amount of responsibility on developers and encourages them to take a satisfying approach, in which they pursue a good-enough solution quickly rather than the optimal one. Indeed, each sprint must finish with the development of a practical and theoretically releasable prototype. Agile advocates for product architectures that enable early component integration, allowing trial versions to be easily reviewed by actual customers even though they have minimal features compared to the final product.

Agile styles, in general, provide developers with a great deal of versatility. Sprint and quick prototyping methods hold minimal product modification costs and marketing time (Cao et al., 2009; Thomke & Reinertsen, 1998). Dynamic project scoping and fast development cycles allow for deferring, making project decisions only when accurate knowledge is available, resulting in an optimum fit between the technological approach and customer requirements.

Hypothesis 5.2
In unpredictable and complex settings, Agile management of new product development systems results in increased quality, speed and cost performance.

Developers tend to mitigate the unpredictability of the environment by using safety margins in the specification and/or design of the new product to allow for variations specific to the project, that is, by overspecification. Developers tend to incorporate just-in-case functionality into product design to protect against changes in consumer tastes or technical trends (Coman & Ronen, 2010).

Researchers regard overspecification of a product outside consumer needs as one of the most significant threats associated with projects of new product development (Coman & Ronen, 2010). Overspecification is expensive, and investment in functionality and performance thresholds that are not required immediately will deplete already-scarce resources allocated to the new product development project (Cao et al., 2009; Bjarnason et al., 2012). The incorporation of external functionalities to satisfy both possible existing and future consumer desires also results in artificially complicated devices that are excessively complex and have little benefit and usefulness to each customer group, a condition known to result in feature fatigue (Thompson et al., 2005; Rust et al., 2006; Buschmann, 2010; Coman & Ronen, 2010; Bjarnason et al., 2012).

Hence, regarding the implications of overspecification, next hypothesis is proposed:

Hypothesis 5.3
Using overspecification in new product development is correlated with decreased quality, speed and higher cost output in unpredictable and competitive settings.

5.3   SAMPLE AND METHODS

The focus of the empirical research in this chapter is on product development, which represents a fruitful sample for research on linear versus non-linear flexible new product development models together with strategies to approach uncertainty during the new product development process.

In terms of ethical considerations for the human subjects in this study, participants' identity remained confidential. They were not asked to provide any personal identifiable information, such as names or addresses (Oldendick, 2012).

A total of 214 product developers in Italy and the United Kingdom were randomly surveyed (Italy 39%; United Kingdom 61%) during the period February 2021–June 2021. The majority of the sampled developers were males (92%) with a minority of females (8%), unlike freelancers they were employed by organisations (71%), they worked on developing products for B2B markets (71%) or B2C markets (29%), and were part of recognised development teams (72%). Most of the participants were aged between 31 and 45 years old (54%) and between 18 and 30 years old (35%) with the residual percentage aged over 45 years old. Their experience in product development exceeded 10 years (46%), between 5 and 10 years (41%), or less than 5 years (13%).

The survey employed multiple-item, seven-point Likert scales to assess the structures of the study's hypotheses, that is, the elements used to evaluate the speed, cost and performance of new product development projects. The dependent variables were based on a substantial body of literature on successful new product development project execution (Kessler & Chakrabarti, 1999; Tatikonda & Rosenthal, 2000). The present scales were grounded in and adapted from the proposal by Marzi (2018). The items and factors used in this analysis are outlined in Table 5.1, together with the mean, factor loadings, composite reliability and average variance extracted. In Table 5.2, "(R)" stands for "reverse" and the items are reverse coded.

Principal component analysis (PCA) with Varimax rotated factors has been used for factor analysis (Hair et al., 2018). The scale items revealed the expected factor variables for speed, cost and quality performance. The factor analysis showed the one-dimensionality of the four-item Stage-Gate factor, which encapsulates its fundamental concepts, including early convergence and model freezing, concurrent development steps and avoidance of late design shifts.

Factor analysis of the Agile and overspecification constructs revealed a three-factor and two-factor solution. In terms of Agile, the three-point-scale sprints encapsulate developers' use of time-boxed, well-defined task cycles to create accurately sized backlog items. Prototype measures the early and regular implementation of prototype experiments and the input they generate. The specified three-item scale evaluates the incremental specification and complex scoping, as well as the overlap between development phases. In terms of overspecification, factor analysis distinguishes between two three-item scales. Extra Product quantifies the inclusion of overspecification designed to create a more performing and "revolutionary" product. Broad Market reflects the developers' desire to provide a product that is inclusive of a diverse customer base.

According to the methodological best-practices of Bagozzi and Yi (1988) together with the suggestions of Steenkamp and Van Trijp (1991), both factor loadings surpass the 0.50 mark, the composite reliability

Table 5.1 Items, loadings and reliability checks

Constructs and measures	Mean (SD)	Standardised factor loadings	Composite reliability	Average variance extracted
Dependent variables				
Speed performance			0.830	0.621
"The majority of the product development projects where I am involved are completed on time"	4.358 (1.645)	0.812		
"In the product development projects where I am involved, milestones and release dates are often postponed compared with the initial plan" (R)	4.151 (1.622)	0.833		
"In the product development projects where I am involved, overwork and time pressure are common in the near-to-launch phases" (R)	3.221 (1.634)	0.714		
Cost performance			0.780	0.541
"The majority of the product development projects where I am involved suffer from budget overruns" (R)	4.132 (1.629)	0.762		
"The cost estimations made at the beginning of the product development projects where I am involved are usually correct"	4.123 (1.637)	0.736		
"In the product development projects where I am involved it often happen that new features or specifications are added without an evaluation of the feasibility and/or without considering the time and resources available" (R)	4.415 (1.632)	0.708		
Quality performance			0.771	0.536
"Compared to other available products, the product development projects where I am involved are more technically cutting-edge"	4.414 (1.332)	0.733		
"The user satisfaction of the product development projects where I am involved is usually high"	5.122 (1.232)	0.874		
"In the product development projects I am involved, malfunctions are often discovered after the product has been launched in the market" (R)	3.222 (1.354)	0.555		

(continued)

Table 5.1 (continued)

Constructs and measures	Mean (SD)	Standardised factor loadings	Composite reliability	Average variance extracted
Independent variables				
Stage-Gate			0.839	0.567
"Shifting product requirements after the project have been already 'frozen' should be categorically avoided"	4.425 (1.647)	0.822		
"When one or more features are added later in the product development projects, they create a distraction from the core features of the new product development projects"	4.223 (1.556)	0.715		
"It is better to 'freeze' the product requirements as early as possible during the product development projects"	4.622 (1.273)	0.722		
"It is better to develop new products only when the product requirements are identified and they are not likely to change in due course"	3.62 (1.67)	0.748		
Agile Factor 1—Sprints			0.798	0.569
"In the product development projects I am involved, products are developed in short reiterated cycles"	4.213 (1.922)	0.772		
"In the product development projects I am involved, the amount of development effort that is completed on each project cycle is clearly defined"	3.935 (1.637)	0.714		
"In the product development projects I am involved, a key feature is typically divided into smaller tasks to fit with the extent of a development cycle"	5.192 (1.579)	0.776		
Agile Factor 2—Prototype testing			0.770	0.527
"It is better to test a prototype version of a new product with users as early as possible, even if it has incomplete functionalities"	5.223 (1.433)	0.742		
"The more frequent the test of prototypes are, the better it is for the product development projects"	4.656 (1.562)	0.715		

(*continued*)

Table 5.1 (continued)

Constructs and measures	Mean (SD)	Standardised factor loadings	Composite reliability	Average variance extracted
"Product requirements should be revised constantly, when new feedback from users is collected after prototype testing"	5.234 (1.432)	0.721		
Agile Factor 3—Specification			0.772	0.530
"It is better to agree product specifications gradually, even during the testing phases of the new product development process"	3.727 (1.667)	0.723		
"It is better to delay product specifications until the beginning of the new product development process"	3.131 (1.544)	0.735		
"When the new products release is distant, it is good practice to include additional features in the current development project"	3.122 (1.343)	0.725		
Overspecification Factor 1—Extra product			0.757	0.510
"To fully satisfy a user, it is important to offer something 'extra', beyond their expectations or needs"	4.233 (1.233)	0.692		
"When specifying the requirements for the new products, it is important to accommodate even the needs of the most advanced users"	4.151 (1.323)	0.683		
"When specifying the requirements for the new products, it is a good practice to include features that are able to anticipate future users' needs and possible technological trajectories"	4.327 (1.532)	0.765		
Overspecification Factor 2—Broad market			0.804	0.578
"To be on the safe side and leave all options open, it is preferable to specify new products requirements broadly and inclusively"	3.734 (1.821)	0.782		

(*continued*)

Table 5.1 (continued)

Constructs and measures	Mean (SD)	Standardised factor loadings	Composite reliability	Average variance extracted
"One key goal in creating new products is to satisfy as many users as possible"	5.112 (1.573)	0.753		
"During the process of developing new products, advanced features can be added even if they are not strictly essential to users"	4.114 (1.532)	0.745		

estimate is greater than 0.60, and the average variance extracted value is greater than 0.50, suggesting that both study models provide a reasonable degree of internal quality.

Maximum Variance Inflation Factors are below the 10-threshold suggested by threshold of 10 recommended by Chatterjee and Hadi (2013). Common method bias has been checked with Harman's Single Factor test and marker variable (Podsakoff et al., 2003).

The following controls were used in the study: age was classified into four categories (18–30 years, 31–45 years, 46–60 years and over 60 years); gender; organisation size was classified into nine categories (single person, 2–4 employees, 5–9, 10–20, 21–40, 41–100, 101–500, 501–1000 and over 1000); the size of the product development team was classified into six categories (independent work, 2–4 members, 5–10, 11–20, 21–40 and greater than 40 members).

The three dependent variables in this analysis were speed, cost and quality efficiency. This study measured Hypotheses 1, 2 and 3 using ordinary least square (OLS) regressions.

5.4   The Effects of Agile and Stage-Gate on Speed, Cost and Quality of the New Product Development Process

Correlations for all variables used in the present study are presented in Table 5.2.

The findings of the OLS regressions are summarised in the next table (Table 5.3). Values in brackets are Standard Errors. Columns 1–3

Table 5.2 Table of correlations among variables

	I	II	III	IV	V	VI	VII	VIII	IX	X	XI	XII	XIII
I Speed													
II Cost	0.531												
III Quality	0.420	0.442											
IV Age	0.032	0.025	0.078										
V Gender	0.045	0.014	0.058	-0.075									
VI Organisation size	0.031	-0.153	-0.041	0.102	0.019								
VII B2C/B2B	0.070	-0.019	-0.055	0.022	0.075	0.055							
VIII Team size	0.088	-0.032	0.071	0.034	-0.122	0.570	-0.096						
IX Stage-Gate	-0.226	-0.140	-0.129	-0.04	-0.142	0.044	-0.021	-0.101					
X Agile—Sprints	0.270	0.228	0.366	-0.061	0.014	0.125	-0.230	0.232	-0.023				
XI Agile—Prototype testing	-0.067	-0.035	-0.030	-0.122	-0.085	-0.057	0.045	0.020	-0.011	0.242			
XII Agile—Specification	0.026	-0.017	-0.027	0.048	0.092	-0.011	-0.027	0.242	-0.127	-0.022	0.071		
XIII Overspecification—Extra product	-0.195	-0.119	0.092	-0.019	-0.023	-0.052	-0.133	0.137	0.190	0.078	0.012	0.210	
XIV Overspecification—Broad market	-0.187	-0.092	0.031	0.073	-0.060	0.166	0.042	-0.015	0.278	0.024	0.038	0.267	0.589

Table 5.3 Results of the OLS regressions

	1	2	3	4	5	6	7	8	9
	Speed 1	*Speed 2*	*Speed 3*	*Cost 1*	*Cost 2*	*Cost 3*	*Quality 1*	*Quality 2*	*Quality 3*
Constant	-0.285	-0.264	-0.196	-0.411	-0.427	-0.577	0.023	0.039	0.222
	(0.549)	(0.521)	(0.523)	(0.534)	(0.565)	(0.536)	(0.554)	(0.554)	(0.556)
Age	-0.042	-0.038	-0.031	0.011	-0.006	0.023	0.055	0.071	0.135
	(0.102)	(0.124)	(0.101)	(0.123)	(0.130)	(0.132)	(0.156)	(0.188)	(0.165)
Gender	-0.025	-0.032	-0.153	-0.234	-0.228	-0.245	0.145	0.045	-0.167
	(0.376)	(0.112)	(0.314)	(0.332)	(0.448)	(0.242)	(0.346)	(0.331)	(0.358)
Organisation size	-0.047	-0.062	-0.021	-0.073*	-0.083*	-0.071*	-0.055	-0.036	-0.055
	(0.032)	(0.045)	(0.079)	(0.021)	(0.040)	(0.022)	(0.023)	(0.025)	(0.065)
B2C/B2B	0.247	0.243	0.256	0.065	0.067	0.046	-0.045	-0.055	-0.031
	(0.169)	(0.163)	(0.233)	(0.173)	(0.171)	(0.155)	(0.166)	(0.166)	(0.097)
Team size	0.049	0.064	0.059	-0.045	-0.047	-0.041	-0.011	0.013	-0.032
	(0.089)	(0.082)	(0.077)	(0.035)	(0.084)	(0.075)	(0.066)	(0.077)	(0.055)
Stage-Gate	-0.329***	-0.366***	-0.313***	-0.233**	-0.232**	-0.231**	-0.144	-0.213*	-0.154*
	(0.065)	(0.073)	(0.066)	(0.057)	(0.088)	(0.069)	(0.067)	(0.079)	(0.056)
Agile—Sprints	0.234***	0.231***	0.266***	0.332***	0.331***	0.357***	0.384***	0.289***	0.466***
	(0.042)	(0.072)	(0.055)	(0.046)	(0.078)	(0.078)	(0.056)	(0.064)	(0.067)
Agile—Prototype testing	-0.173*	-0.153*	-0.123	-0.132	-0.214	-0.112	-0.111	-0.077	-0.045
	(0.075)	(0.073)	(0.033)	(0.069)	(0.078)	(0.065)	(0.074)	(0.045)	(0.023)
Agile—Specification	-0.076	-0.071	-0.111	-0.061	-0.061	-0.086	-0.047	-0.064	-0.099
	(0.079)	(0.072)	(0.076)	(0.083)	(0.087)	(0.079)	(0.069)	(0.067)	(0.056)
Overspecification—Extra product	-0.211**	-0.222**	-0.183**	-0.132*	-0.189*	-0.177	0.092	0.214	0.098
	(0.073)	(0.075)	(0.072)	(0.056)	(0.081)	(0.032)	(0.085)	(0.023)	(0.056)
Overspecification—Broad market	-0.173***	-0.219***	-0.212***	-0.123*	-0.147*	-0.066*	0.031	0.045	0.077
	(0.097)	(0.081)	(0.069)	(0.071)	(0.084)	(0.076)	(0.077)	(0.055)	(0.069)

	(1)	(2)	(3)	(4)	(5)	(6)
Stage-Gate × sprints	0.124* (0.063)		0.085 (0.082)		0.211* (0.075)	
Stage-Gate × prototype testing	−0.034 (0.623)		−0.047 (0.064)		0.039 (0.056)	
Stage-Gate × Agile—specification	−0.127* (0.062)		−0.023 (0.068)		0.011 (0.064)	
Sprints × prototype testing		0.034 (0.044)		−0.031 (0.098)		−0.055 (0.064)
Sprints × Agile –specification		0.031 (0.073)		0.044 (0.065)		0.177* (0.034)
Prototype testing × Agile—Specification		0.171* (0.074)		0.085 (0.065)		0.139 (0.076)
Adjusted R^2	0.222	0.195	0.185	0.165	0.197	0.262
	0.243	0.223	0.169			

* $p \leq 0.05$; ** $p \leq 0.01$; *** $p \leq 0.001$

correspond to the dependent variable, speed; columns 4–6 to the dependent variable, cost; and columns 7–9 to the dependent variable, quality. According to Hypothesis 5.1, adhering to management concepts compatible with the Stage-Gate method resulted in lower quality, speed and cost output. Since the Stage-Gate coefficients were negative and statistically consistent across models, Hypothesis 5.1 is accepted.

Hypothesis 5.2 assumed a correlation between Agile approach to the new product development process and performance. The results are mixed: sprinting is positively associated with increased performance on speed, cost and quality for all models; in columns 1–3, the prototype coefficient is negative and significant, implying that premature and recurrent prototype testing is related to decreased speed performance.

A negative relationship between overspecification and speed performance is predicted by Hypothesis 5.3. Overspecification coefficients are negative and relevant for speed performance, for both extra product and broad market approach. Overspecification also shows a weak, yet significant negative correlation with cost. These findings support Hypothesis 5.3.

The coefficients of the relationship term between Stage-Gate and Sprints are positive and significant in columns 2 and 8, implying that organising development in working cycles mitigates Stage-Gate's possible negative impact on speed and quality performance. In Column 2, the interaction between Stage-Gate and Agile-Specification is negative and significant, which means that following Stage-Gate while adhering to the Agile concepts of incremental development and dynamic scoping results in decreased speed performance.

This analysis conducts a post-hoc examination to determine the complementarity of the three Agile variables found by factor analysis. The coefficients of the interaction among Prototype Testing and Agile-Specification are positive and significant in Column 3, showing that combining these two Agile principles results in faster projects. The coefficient of the interaction between Sprint and Agile-Specification is also positive and significant in Column 9, highlighting that they have a beneficial impact on quality performance.

This chapter addresses the implications of innovation management and research of new product development. The empirical research shows that attempting to excessively control uncertainty by wide-ranging and upfront planning could cause time and cost overruns. This finding is consistent with previous new product development research (Antons et al., 2019; Cao et al., 2009; Thomke, 1997; Ward & Sobek II, 2014; Sommer et al., 2015).

This study also provides a new and complex interpretation of Agile models and the impact they have on developing a new product. The information gathered from the survey, which is used in this chapter, indicates that several distinct Agile components are present, and developers implement each of the components in a different way and each component has a diverse effect on project outcomes.

Furthermore, the present study examines differences in the relationship between the Agile principles and the performance of new product development. The study demonstrates that using sprints generates improved time-to-market, cost-to-market and fit-to-market. This shows the benefits of organising development work in brief, explicit iterative loops and contributes to the increasing amount of research in more than just the Agile area (Antons et al., 2019; Ries, 2011).

This research indicates that overspecification approach has several negative correlations with the speed. In fact, developing additional features requires additional time. Also, developing an overspecified product leads to additional costs, as shown by the negative correlation with cost performance. It is not yet understood if the time (and the resources) dedicated to overspecify a product negatively influenced the quality of the other functionalities (Buschmann, 2010; Coman & Ronen, 2010). It must also be taken into account the effect of overspecified products on the final users, resulting in possible feature fatigue (Thompson et al., 2005).

This study also suggests some practical recommendations. Managers, innovation managers, entrepreneurs and start-uppers who work in environments that are volatile will discover that breaking the new development process into several sprints improves the success of the project.

Linear, plan-based Stage-Gate models would need avoiding in such situations due to their possible inability to deal with the complexity inherent in creative endeavours. Furthermore, managers, innovation managers, entrepreneurs and start-uppers are cautioned that unnecessary prototype testing can result in schedule slippage without advancing product-market fit.

Many shortcomings exist in the present research, requiring additional effort to extend it. The present study employs a retrospective approach from a single respondent, however, that is the approach expected in wide-ranging survey research on new product development projects. As a result, retrospective and methodological biases can occur. This research examines the product development industry, which is unique compared to other industries. While conceptualising Stage-Gate, Agile and overspecification in a dimension of perceptions rather than tools can increase the possibility of generalising the results, further studies ought to validate the findings examined in this research across countries and industries.

Future research can incorporate information from a diversity of industries to project environments and investigate how the importance of various management concepts shifts as ambiguity, volatility and complexity increase. The study does not include critical elements, such as dedicated cross-functional teams, because they are not specifically linked to managing uncertainty. Further research should provide a better, detailed and more systematic evaluation of Agile components, the interaction of the components and the effect they have on how the new product development projects perform.

REFERENCES

Antons, D., Brettel, M., Hopp, C., Salge, T. O., Piller, F., & Wentzel, D. (2019). Stage-gate and agile development in the digital age: Promises, perils, and boundary conditions. *Journal of Business Research, 110*, 495–501.

Bagozzi, R. P., & Yi, Y. (1988). On the evaluation of structural equation models. *Journal of the Academy of Marketing Science, 16*(1), 74–94.

Bayus, B. L. (2013). Crowdsourcing new product ideas over time: An analysis of the Dell IdeaStorm community. *Management Science, 59*(1), 226–244.

Beck, K., Beedle, M., Van Bennekum, A., Cockburn, A., Cunningham, W., Fowler, M., & Kern, J. (2001). The agile manifesto.

Birkinshaw, J. (2018). What to expect from agile. *MIT Sloan Management Review, 59*(2), 39–42.

Bjarnason, E., Wnuk, K., & Regnell, B. (2012). Are you biting off more than you can chew? A case study on causes and effects of overscoping in large-scale software engineering. *Information and Software Technology, 54*(10), 1107–1124.

Boehm, B. W. (1991). Software risk management: Principles and practices. *IEEE Software, 8*(1), 32–41.

Buschmann, F. (2010). Learning from failure, part 2: Featuritis, performitis, and other diseases. *IEEE Software, 27*(1), 10–11.

Cao, L., Mohan, K., Xu, P., & Ramesh, B. (2009). A framework for adapting agile development methodologies. *European Journal of Information Systems, 18*(4), 332–343.

Chatterjee, S., & Hadi, A. S. (2013). *Regression analysis by example*. John Wiley & Sons.

Choi, D., & Valikangas, L. (2001). Patterns of strategy innovation. *European Management Journal, 19*(4), 424–429.

Coman, A., & Ronen, B. (2010). Icarus' predicament: Managing the pathologies of overspecification and overdesign. *International Journal of Project Management, 28*(3), 237–244.

Cooper, R. G. (2014). What's next? After stage-gate. *Research-Technology Management, 57*(1), 20–31.

Cooper, R. G., Edgett, S. J., & Kleinschmidt, E. J. (2002). Optimizing the stage-gate process: What best-practice companies do—II. *Research-Technology Management, 45*(6), 43–49.

Cooper, R. G., & Sommer, A. F. (2016). The agile–stage-gate hybrid model: a promising new approach and a new research opportunity. *Journal of Product Innovation Management, 33*(5), 513–526.

Ernst, H. (2002). Success factors of new product development: A review of the empirical literature. *International Journal of Management Reviews, 4*(1), 1–40.

Ettlie, J. E., & Elsenbach, J. M. (2007). Modified stage-gate® regimes in new product development. *Journal of Product Innovation Management, 24*(1), 20–33.

Ettlie, J. E., & Rosenthal, S. R. (2011). Service versus manufacturing innovation. *Journal of Product Innovation Management, 28*(2), 285–299.

Evanschitzky, H., Eisend, M., Calantone, R. J., & Jiang, Y. (2012). Success factors of product innovation: An updated meta-analysis. *Journal of Product Innovation Management, 29*(S1), 21–37.

Hair, J. B., Black, W., Anderson, R., & Babin, B. J. (2018). *Multivariate Data Analysis* (8th ed.). Cengage Learning EMEA.

Iansiti, M. (1995). Shooting the rapids: Managing product development in turbulent environments. *California Management Review, 38*(1), 37–58.

Kahneman, D., & Tversky, A. (1977). *Intuitive prediction: Biases and corrective procedures.* Decisions and Designs.

Karlsson, C., & Ahlström, P. (1996). The difficult path to lean product development. *Journal of Product Innovation Management, 13*(4), 283–295.

Karlstrom, D., & Runeson, P. (2005). Combining agile methods with stage-gate project management. *IEEE Software, 22*(3), 43–49.

Kessler, E. H., & Chakrabarti, A. K. (1999). Speeding up the pace of new product development. *Journal of Product Innovation Management, 16*(3), 231–247.

Krishnan, V., Eppinger, S. D., & Whitney, D. E. (1997). A model-based framework to overlap product development activities. *Management Science, 43*(4), 437–451.

Krishnan, V., & Zhu, W. (2006). Designing a family of development-intensive products. *Management Science, 52*(6), 813–825.

Lee, H., & Markham, S. K. (2016). PDMA comparative performance assessment study (CPAS): Methods and future research directions. *Journal of Product Innovation Management, 33*(S1), 3–19.

Liker, J. K., & Morgan, J. M. (2006). The Toyota way in services: The case of lean product development. *The Academy of Management Perspectives, 20*(2), 5–20.

Loch, C. H., DeMeyer, A., & Pich, M. (2011). *Managing the unknown: A new approach to managing high uncertainty and risk in projects.* John Wiley & Sons.

Marzi, G. (2018). *Product and process innovation: From manufacturing to IT firms.* Doctoral dissertation, University of Pisa. Available from ETD database.

Marzi, G. (2022). On the nature, origins and outcomes of Over Featuring in the new product development process. *Journal of Engineering and Technology Management, 64,* 101–685

Michaelis, T. L., Aladin, R., & Pollack, J. M. (2018). Innovation culture and the performance of new product launches: A global study. *Journal of Business Venturing Insights, 9,* 116–127.

Montoya-Weiss, M. M., & Calantone, R. (1994). Determinants of new product performance: A review and meta-analysis. *Journal of Product Innovation Management, 11*(5), 397–417.

Oldendick, R. W. (2012). *Survey research ethics. In handbook of survey methodology for the social sciences* (pp. 23–35). Springer.

Podsakoff, P. M., MacKenzie, S. B., Lee, J. Y., & Podsakoff, N. P. (2003). Common method biases in Behavioral research: A critical review of the literature and recommended remedies. *Journal of Applied Psychology, 88*(5), 879–903.

Ries, E. (2011). *The lean startup: How today's entrepreneurs use continuous innovation to create radically successful businesses.* Crown Business.

Ronen, B., & Pass, S. (2008). *Focused operations management: Achieving more with existing resources.* John Wiley & Sons.

Rust, R. T., Thompson, D. V., & Hamilton, R. W. (2006). Defeating feature fatigue. *Harvard Business Review, 84*(2), 37–47.

Schreier, M., Fuchs, C., & Dahl, D. W. (2012). The innovation effect of user design: Exploring consumers' innovation perceptions of firms selling products designed by users. *Journal of Marketing, 76*(5), 18–32.

Sommer, A. F., Hedegaard, C., Dukovska-Popovska, I., & Steger-Jensen, K. (2015). Improved product development performance through agile/stage-gate hybrids: The next-generation stage-gate process? *Research-Technology Management, 58*(1), 34–45.

Steenkamp, J. B. E., & Van Trijp, H. C. (1991). The use of LISREL in validating marketing constructs. *International Journal of Research in Marketing, 8*(4), 283–299.

Sull, D. N. (2004). Disciplined entrepreneurship. *MIT Sloan Management Review, 46*(1), 71.

Tatikonda, M. V., & Rosenthal, S. R. (2000). Technology novelty, project complexity, and product development project execution success: A deeper look at task uncertainty in product innovation. *IEEE Transactions on Engineering Management, 47*(1), 74–87.

Terwiesch, C., & Ulrich, K. T. (2009). *Innovation tournaments: Creating and selecting exceptional opportunities.* Harvard Business Press.

Thomke, S., & Reinertsen, D. (1998). Agile product development: Managing development flexibility in uncertain environments. *California Management Review, 41*(1), 8–30.

Thomke, S. H. (1997). The role of flexibility in the development of new products: An empirical study. *Research Policy, 26*(1), 105–119.

Thompson, D. V., Hamilton, R. W., & Rust, R. T. (2005). Feature fatigue: When product capabilities become too much of a good thing. *Journal of Marketing Research, 42*(4), 431–442.

Ward, A. C., & Sobek, D. K., II. (2014). *Lean product and process development.* Lean Enterprise Institute.

CHAPTER 6

The Threat of Uncertainty in the New Product Development Process

Abstract Entrepreneurs, innovators and managers are asked to consider where and how they faced unnecessary development issues in their daily work, emphasising the importance of specifying the boundaries of a new product development project and how an unnecessary load of features would affect the fulfilment of a development project. Several project managers' and developers' perceptions, underestimation of a project's complexity and a blurred understanding of what consumers want are among the causes of unsustainable product development. Entrepreneurs, innovators and managers should consider the value drivers of a product, as well as the dangers of an overly developed product. The aim of an effective new product development project should be to maximise market positioning while preserving product and usability.

Keywords New product development • Uncertainty • Excessive development

G. Marzi, *Uncertainty-driven Innovation*,
https://doi.org/10.1007/978-3-030-99534-8_6

6.1   THE DETRIMENTAL CONDITION OF THE NEW PRODUCT DEVELOPMENT PROCESS IN A HIGH UNCERTAINTY ENVIRONMENT

The Over Featuring forms, under two of the most popular new product development project management tools, Stage-Gate and Agile, were only partly discussed in the available theoretical and empirical literature (Cooper & Edgett, 2008; Karlstrom & Runeson, 2005). It is important to assess when, why and how Over Featuring forms are likely to occur during the particular phases of the Stage-Gate or Agile approaches of creating a new product, as shown in the previous chapter of this book.

Dangerous outcomes can be overscoping and overspecification, which should be avoided at the beginning of developing a new product, whereas the other Over Featuring forms may be difficult to control and classify since they seem to appear all at once later in the process (Marzi, 2022). However, potential changes to the requirements or the nature of the new product development project pose an added danger during development, testing and review, as well as post-launch.

Over Featuring forms are expected to appear in almost the same stages of the Stage-Gate method as they appear when using the Agile method during the development process of a new product. The phases of development are unpacked and replicated many times in an Agile-based new product development process following ongoing checks and reviews with partners and end-users (Bianchi et al., 2020; Birkinshaw, 2018). The Agile method gives the project more stability, makes for a more incremental understanding of the scope and specifications, and allows for constant revision of the functionality used in the product's different updates. However, Agile's extreme versatility may be a double-edged sword for excessive development (Bianchi et al., 2020; Schreier et al., 2012). Surprisingly, the conditions that favour the emergence of Over Featuring are better than those that favour Stage-Gate (Antons et al., 2019; Bjarnason et al., 2012; Schmidt & Calantone, 2002). The literature showed that the ongoing updates of the scope and specifications, as well as the rapid and incessant implementation cycles, could subject the whole new product development project to an additional escalation of excessive development when compared to Stage-Gate (Antons et al., 2019; Bianchi et al., 2019, 2020).

However, the Stage-Gate-Hybrid strategy might be a viable solution to reduce the drawbacks of Over Featuring (Bianchi et al., 2020).

Stage-Gate-Hybrid may be a promising alternative to reducing the negative effects of Over Featuring during the new product development process since it blends the reliability of a linear process with the constant availability of inputs from the different stakeholders taking part in the creation process of a novel product. While a purely hybrid solution is unlikely to be appropriate, it provides a sound basis for implementing excessive development–lowering activities (Marzi, 2022).

As seen in the previous chapters, Over Featuring tend to manifest at different phases in the action of establishing a new product and frequently they are not decoupled from one another, but are strongly intertwined. Over Featuring is related to excessive development through the entire New Product Development process (Marzi, 2022). Figure 6.1 shows the logical flow of the new product development process where the excessive development can manifest.

Uncertainty, market, cognitive and project-driven antecedents can be identified as the four major driving factors for over featuring (Marzi, 2022). The first group, uncertainty-driven, encompasses all situations in which the external environment's acute complexity drives the new product development project beyond the margin of tolerance. This may arise, for example, if the project's fuzzy front-end cannot be mitigated or is not adequately mitigated (Antons et al., 2019; Bianchi et al., 2019). In this circumstance, broadening the scope, tightening the specifications or inserting additional functionality provides a buffer against the ambiguity of the customers' necessities. Excessive safety margins against the volatility of the outer environment, on the other hand, may be a source of Over Featuring (Bjarnason et al., 2012). Engineers, innovators, R&D directors and project managers naturally have backups to fix potential future product upgrades if the desires of the customers or the product's usage cannot be easily understood throughout the new product development project. Similarly, whether the technical acceleration is swift or the regulatory

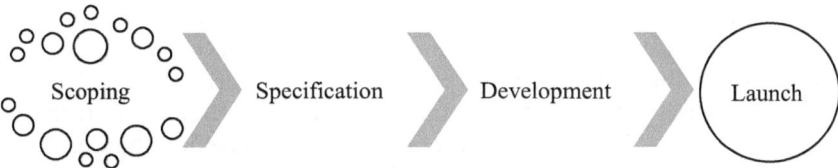

Fig. 6.1 The new product development stages

climate is turbulent, the addition of extra and unneeded functionality during the project of a new product creation is likely going to arise for the purpose of foreseeing, as well as resolving potential consumer or industry shifts (Coman & Ronen, 2010; Michaelis et al., 2018).

From the standpoint of market-driven antecedents, research has shown that marketing and technological departments often compete to achieve a balance between strategic viability and consumer appeal (Choi & Valikangas, 2001; Michaelis et al., 2018; Rust et al., 2006). The "leave-all-options-open strategy" and the "one-size-fits-all approach" are two of the more prominent antecedents of Over Featuring. The first is a deliberate delay in defining project limits to properly react to changing consumer needs (Antons et al., 2019; Bianchi et al., 2019). The second is the erroneous assumption that consumers want an all-in-one package that meets their needs (Thompson et al., 2005). Nonetheless, to draw new or returning buyers, marketing teams often request unreasonable functionality or results (Thompson et al., 2005). Similarly, undue acquiescence to users' wishes may lead to the project's incorporation of qualitatively low features or special features with a small user base (Buschmann, 2010). Finally, competitive demand is a significant factor in the generation of excessive development. The competitive strain focused on features is especially important in several mass consumer industries, such as domestic appliances (Rust et al., 2006). Features are seen as a marketing advantage to win potential buyers, boost repurchases and keep existing ones (Karlstrom & Runeson, 2005; Thompson et al., 2005). As a result, top management's pressure on marketing and product teams to incorporate new and improved functionality is significant (Karlsson & Ahlström, 1996; Rust et al., 2006).

In terms of cognitive forces, a variety of research has shown that project managers, technicians, developers and R&D managers' human cognitive biases, cognitive types, desires and habits have a significant effect on the generation of Over Featuring. Overconfidence, anchoring, sunk-cost and perfectionism are all examples of human bias that play a part in decision-making throughout the process of a new product creation. A mechanism for making skewed decisions encourages the unchecked expansion of functionality, according to the literature, because developers appear to overestimate the utility of their own designed features (Shmueli et al., 2016). Similarly, emotional commitment to a mission, a certain technology or a certain collection of features causes a chain reaction that results in excessive development (Bianchi et al., 2019).

When it comes to project-driven antecedents, excessive development's pursuit of performance improvement may be a double-edged weapon. Involved parties, such as managers dealing with projects, engineers and others during the process of developing a new product, are often pressed to use the most up-to-date or best-in-class technologies in their products (Karlsson & Ahlström, 1996; Davis & Venkatesh, 2004). When a revolutionary technology arrives on the market and the production team pushes to incorporate it into the project, looking for quality improvement is especially risky. A transition in the design of the product because it includes modern technologies, if not well handled, may be a powerful antecedent for excessive development (Choi & Valikangas, 2001; Davis & Venkatesh, 2004). As a result, particularly during the late stages of development, a suitable compromise between the advantages of emerging technologies and the drawbacks of an untested technology should be contemplated.

Similarly, during both the development and specification phases, an unregulated and persistent stream of requirements inflow is a trigger of excessive development. In this situation, if project managers do not serve as a filter between the market's requests and the implementation teams' operations by reducing requirement inflows, the probability of excessive development rises in proportion to the amount of new requirement requests. Having an unclear overview of the allocated resources to the project creates a propensity to overshoot the project's scale and intricacy and an ambitious project overview from the perspective of resources and objectives.

Unfocused project priorities are also precursors to scope and feature bloat since the project's limits are hazy (Bjarnason et al., 2012). The planning team's lack of participation during the specification process results in a misalignment between the technological viability of specific project features and the needs of consumers (Coman & Ronen, 2010). To prevent the introduction of impractical features proposed to consumers by marketing or sales departments, technical teams must be active in the specification process (Bjarnason et al., 2012).

6.2   The Effects of Excessive Development on the New Product Development Process

Once excessive development have been pinpointed and disseminated across the new product development project, a sequence of signs appears as an expression of the underlying Over Featuring.

Thompson et al. (2005) described feature fatigue as the frustration users feel while utilising products with a large and overwhelming range of features. Feature fatigue is concerned with the amount of product capabilities: since there are so many, consumers suffer many negative consequences as an effect of using the product daily (Choi & Valikangas, 2001; Rust et al., 2006).

Thompson et al. (2005) showed that the number of features drives three desired outcomes. The goal is to optimise consumers' net present value by providing a product with a well-balanced range of features. The second is to increase initial purchasing by providing an appealing yet bloated product. The third is to increase repurchase rates by providing a basic but consistent product with a small set of features. As a consequence, depending on the intended goal, companies can set the product's limits between the second and third points. Thompson et al. (2005), therefore, explained the impact of features from the customer's perspective and the organisation's deliberate approach.

Though Thompson et al. (2005) found a connection between feature fatigue and bloat in features, the existing literature largely ignored the link between Over Featuring and feature fatigue. Looking at the available studies, it is apparent that feature fatigue is one of the more common normal outcomes of excessive development, even though it is rarely explicitly mentioned (De Giovanni, 2019; Wu et al., 2015). Thompson et al. (2005) empirically discovered that one reason for including an excessive number of features can be traced to pressure from the marketing department during the new product development phase to have further functionalities, similar to other forms of Over Featuring, including overspecification, overscoping and feature creep (Bjarnason et al., 2012; Coman & Ronen, 2010; Elliott, 2007; Jain, 2019). Even feature fatigue has been shown in the literature as a terminal and tangible consequence stemming from a more nuanced underlying state described by Over Featuring (Marzi, 2022).

Regarding the quality issues and unmet customer expectations resulting from excessive development, the literature has revealed a strong link between excessive development and these issues (Birkinshaw, 2018; Bjarnason et al., 2012; Elliott, 2007). The fundamental human tendency to think about any initiative as sequential ignores the fact that each new function applied to the project increases the sophistication of the system's design exponentially. When new enhancements are introduced, they begin to fight for the resources of the project in connection with financial resources, time and qualitative focus. Increasing workloads resulting from continuous specifications and feature inflow diverts the project team's resources and energy away from quality and customer needs (Bjarnason et al., 2012). As a consequence, products are often shipped with a certain low-product functionality, resulting in malfunctioning or defective features that fail to satisfy end-user expectations (Bjarnason et al., 2012; Davis & Venkatesh, 2004). As a result, bloat in functionality affects feature fatigue as well as the overall project's consistency where a new product is developed, resulting in a very quick exacerbation of negative results (Buschmann, 2010; Gill, 2008; Schmidt & Calantone, 2002).

Excessive development effects are not restricted to the launch and after-launch stages, as shown by the literature. Several signs can appear while new products are being created, leading up to the actual launch. Budget overruns, project delays and a lack of care are all visible, as excessive development disperse and infiltrate the new product development project, particularly in the final stages of development before launch (Birkinshaw, 2018; Garcia et al., 2019; Schmidt & Calantone, 2002). The excitement generated by a powerful and feature-rich product will soon collide with the considerable refactoring required to complete, implement and evaluate many features, which will almost certainly result in enormous costs and schedule delays (Buschmann, 2010). New enhancements battle for the assets allotted to the project; more features equate to less time dedicated to each aspect and higher overall project costs.

It is also imperative to figure out how and to what degree external, uncontrollable factors like market volatility and technical progress play a role in excessive development generation. To determine the force of outer factors on the emergence of excessive development within organisations, a large number of new product development ventures should be investigated. While there is recognition of certain internal antecedents, future research is required to merge the obtained results and investigate fresh, unknown antecedents.

While some research has looked into the effects of excessive development on product and project success, continued research is necessary to unpack and clarify what influence excessive development directly have on particular aspects of the efficiency of developing a new product, such as prices, time, consistency, consumer expectation and product placement. However, it is still unclear where, how and to what extent excessive development impacts new product development output and, by implication, the company's performance throughout the project (Garcia et al., 2019).

Although the literature alludes to a widespread correlation between excessive development and feature fatigue, only a few experiments empirically related the two. The role of features in product performance and positioning is critical, as Thompson et al. (2005) demonstrated in terms of the company's product range and overall strategy.

A key concern with the unresolved study query for excessive development is the lack of a multidimensional estimation method that includes at least the major forms of Over Featuring. After the creation of a robust calculation method, the natural evolution could also lead to the development of an assessment tool, perhaps driven by the Balanced Scorecard approach.

Preventive steps can therefore be taken during the process of new product creation to stop excessive development from occurring.

References

Antons, D., Brettel, M., Hopp, C., Salge, T. O., Piller, F., & Wentzel, D. (2019). Stage-gate and agile development in the digital age: Promises, perils, and boundary conditions. *Journal of Business Research, 110*, 495–501.

Bianchi, M., Marzi, G., & Guerini, M. (2020). Agile, stage-gate and their combination: Exploring how they relate to performance in software development. *Journal of Business Research, 110*, 538–553.

Bianchi, M., Marzi, G., Zollo, L., & Patrucco, A. (2019). Developing software beyond customer needs and plans: An exploratory study of its forms and individual-level drivers. *International Journal of Production Research, 57*(22), 7189–7208.

Birkinshaw, J. (2018). What to expect from agile. *MIT Sloan Management Review, 59*(2), 39–42.

Bjarnason, E., Wnuk, K., & Regnell, B. (2012). Are you biting off more than you can chew? A case study on causes and effects of overscoping in large-scale software engineering. *Information and Software Technology, 54*(10), 1107–1124.

Buschmann, F. (2010). Learning from failure, part 2: Featuritis, performitis, and other diseases. *IEEE Software, 27*(1), 10–11.

Choi, D., & Valikangas, L. (2001). Patterns of strategy innovation. *European Management Journal, 19*(4), 424–429.

Coman, A., & Ronen, B. (2010). Icarus' predicament: Managing the pathologies of overspecification and overdesign. *International Journal of Project Management, 28*(3), 237–244.

Cooper, R. G., & Edgett, S. J. (2008). Maximizing productivity in product innovation. *Research-Technology Management, 51*(2), 47–58.

Davis, F. D., & Venkatesh, V. (2004). Toward preprototype user acceptance testing of new information systems: Implications for software project management. *IEEE Transactions on Engineering Management, 51*(1), 31–46.

De Giovanni, P. (2019). A feature fatigue supply chain game with cooperative programs and ad-hoc facilitators. *International Journal of Production Research, 57*(13), 4166–4186.

Elliott, B. (2007). Anything is possible: Managing feature creep in an innovation rich environment. *IEEE International Engineering Management Conference, 2007*, 304–307.

Garcia, J. J., Pettersen, S. S., Rehn, C. F., Erikstad, S. O., Brett, P. O., & Asbjørnslett, B. E. (2019). Overspecified vessel design solutions in multi-stakeholder design problems. *Research in Engineering Design, 30*(4), 473–487.

Gill, T. (2008). Convergent products: What functionalities add more value to the base? *Journal of Marketing, 72*(2), 46–62.

Jain, S. (2019). Time inconsistency and product design: A strategic analysis of feature creep. *Marketing Science, 38*(5), 835–851.

Karlsson, C., & Ahlström, P. (1996). The difficult path to lean product development. *Journal of Product Innovation Management, 13*(4), 283–295.

Karlstrom, D., & Runeson, P. (2005). Combining agile methods with stage-gate project management. *IEEE Software, 22*(3), 43–49.

Marzi, G. (2022). On the nature, origins and outcomes of Over Featuring in the new product development process. *Journal of Engineering and Technology Management, 64*, 101–685.

Michaelis, T. L., Aladin, R., & Pollack, J. M. (2018). Innovation culture and the performance of new product launches: A global study. *Journal of Business Venturing Insights, 9*, 116–127.

Rust, R. T., Thompson, D. V., & Hamilton, R. W. (2006). Defeating feature fatigue. *Harvard Business Review, 84*(2), 37–47.

Schmidt, J. B., & Calantone, R. J. (2002). Escalation of commitment during new product development. *Journal of the Academy of Marketing Science, 30*(2), 103–118.

Schreier, M., Fuchs, C., & Dahl, D. W. (2012). The innovation effect of user design: Exploring consumers' innovation perceptions of firms selling products designed by users. *Journal of Marketing, 76*(5), 18–32.

Shmueli, O., Pliskin, N., & Fink, L. (2016). Can the outside-view approach improve planning decisions in software development projects? *Information Systems Journal, 26*(4), 395–418.

Thompson, D. V., Hamilton, R. W., & Rust, R. T. (2005). Feature fatigue: When product capabilities become too much of a good thing. *Journal of Marketing Research, 42*(4), 431–442.

Wu, M., Wang, L., Long, H., & Li, M. (2015). Feature fatigue analysis in product development. *Total Quality Management and Business Excellence, 26*(1–2), 218–232.

CHAPTER 7

Conclusion

Abstract The analytical findings, along with an examination of the current theory, confirmed the critical role of uncertainty in influencing innovation in companies. In particular, innovating a product and a process, along with the creation of a completely new product, is highly affected by the unpredictability of the external environment. Entrepreneurs, managers, innovators and start-uppers have only limited tools to mitigate the adverse effects of an always changing external environment. The present chapter summarises the findings of the present book and proposes a springboard to further explore this pivotal topic of research.

Keywords Innovation • Product • Agile • Excessive development • Overspecification

The present book first demonstrates the broader field of analysis of innovating processes and products in companies in the manufacturing industry, showing that various research sub-directions within such an area of study are available, as other scholars have illustrated (Damanpour, 1991; Garcia & Calantone, 2002; Marzi et al., 2021).

The analysis of the second topic shows how process innovation could have a wide-ranging impact within companies, contributing to the

© The Author(s), under exclusive license to Springer Nature   105
Switzerland AG 2022
G. Marzi, *Uncertainty-driven Innovation*,
https://doi.org/10.1007/978-3-030-99534-8_7

emergence of innovative products (Reichstein & Salter, 2006), as well as strategic advantage and long-term company sustainability, confirmed by an empirical demonstration with a case study (Damanpour & Gopalakrishnan, 2001; Martinez-Ros, 1999; Marzi et al., 2018; Sirilli & Evangelista, 1998).

Next, the study concentrates on some potentially negative effects of excessive development and Over Featuring on new product development (Marzi, 2022).

Proposing an analytical enquiry into the significance of uncertainty and excessive development in the development process of a new product, the present book emphasises how complexity affects product innovation and growth. Several approaches, such as Agile and Stage-Gate, have been created to overcome this kind of challenge, with varying outcomes (Cooper & Sommer, 2016; Sommer et al., 2015). Next, the book places an emphasis on the role of excessive development in the various stages and methods of developing a new product, as well as the Stage-Gate and Agile frameworks showing how they have the potential to ease excessive development.

However, in rapidly evolving business conditions, developing a product poses a continuous task owing to growing complexity and fluctuating consumer demands (Karlsson & Ahlström, 1996). As a result, particularly in the late stages of the development process, developing a new product also causes many iterations of trials and evaluations, including constant revisions to satisfy standards and specifications of the market (Birkinshaw, 2018; MacCormack & Verganti, 2003).

In this regard, a common approach to complexity is the norm, and maybe implicit, with a propensity to add increasing amounts of functionality to the new product in the expectation of meeting the demands of a broader range of consumers. This reaction to ambiguity is known as excessive development as already shown.

Excessive development strategies are used to build a "safe margin" throughout the process of creating a new product to broaden the spectrum of consumer demands met by the product and create a barrier against the uncertainty. One possible way is the overspecification discussed in the third section of this book.

As a result, excessive development configure in drawbacks driven by uncertainty. On the one hand, entrepreneurs, innovators, developers and R&D managers want to provide a buffer to deal with potential and unpredictable consumer needs, on the other hand, however, the deceptive cognitive understanding of their projects can contribute to an overall warped perception of the scheme (Bianchi et al., 2019).

Despite the widespread diffusion of excessive development, scholars have not placed enough emphasis on the subject, especially the roots, results and measurements. Currently, only a few studies have looked at excessive development in detail, with results that are frequently contentious and fragmented. With this in mind, theoretical as well as empirical research is required to investigate the antecedents, limits and implications of excessive development and their role in uncertainty.

Thus, potential study opportunities are available: while the theoretical perspectives offered by previously reviewed studies in this book need to be enriched, there is a need for additional empirical and theoretical material to ground further development of the field.

REFERENCES

Bianchi, M., Marzi, G., Zollo, L., & Patrucco, A. (2019). Developing software beyond customer needs and plans: An exploratory study of its forms and individual-level drivers. *International Journal of Production Research, 57*(22), 7189–7208.

Birkinshaw, J. (2018). What to expect from agile. *MIT Sloan Management Review, 59*(2), 39–42.

Cooper, R. G., & Sommer, A. F. (2016). The agile-stage-gate hybrid model: A promising new approach and a new research opportunity. *Journal of Product Innovation Management, 33*(5), 513–526.

Damanpour, F. (1991). Organizational innovation: A meta-analysis of effects of determinants and moderators. *Academy of Management Journal, 34*(3), 555–590.

Damanpour, F., & Gopalakrishnan, S. (2001). The dynamics of the adoption of product and process innovations in organizations. *Journal of Management Studies, 38*(1), 45–65.

Garcia, R., & Calantone, R. (2002). A critical look at technological innovation typology and innovativeness terminology: A literature review. *Journal of Product Innovation Management, 19*(2), 110–132.

Karlsson, C., & Ahlström, P. (1996). The difficult path to lean product development. *Journal of Product Innovation Management, 13*(4), 283–295.

MacCormack, A., & Verganti, R. (2003). Managing the sources of uncertainty: Matching process and context in software development. *Journal of Product Innovation Management, 20*(3), 217–232.

Martinez-Ros, E. (1999). Explaining the decisions to carry out product and process innovations: The Spanish case. *The Journal of High Technology Management Research, 10*(2), 223–242.

Marzi, G. (2022). On the nature, origins and outcomes of Over Featuring in the new product development process. *Journal of Engineering and Technology Management, 64*, 101–685.

Marzi, G., Ciampi, F., Dalli, D., & Dabic, M. (2021). New product development during the last ten years: The ongoing debate and future avenues. *IEEE Transactions on Engineering Management, 68*(1), 330–344.

Marzi, G., Zollo, L., Boccardi, A., & Ciappei, C. (2018). Additive manufacturing in SMEs: Empirical evidences from Italy. *International Journal of Innovation and Technology Management, 15*(01), 1850007.

Reichstein, T., & Salter, A. (2006). Investigating the sources of process innovation among UK manufacturing firms. *Industrial and Corporate Change, 15*(4), 653–682.

Sirilli, G., & Evangelista, R. (1998). Technological innovation in services and manufacturing: Results from Italian surveys. *Research Policy, 27*(9), 881–899.

Sommer, A. F., Hedegaard, C., Dukovska-Popovska, I., & Steger-Jensen, K. (2015). Improved product development performance through agile/stage-gate hybrids: The next-generation stage-gate process? *Research-Technology Management, 58*(1), 34–45.

INDEX